MINARET BUILDING

AND APPRENTICESHIP

IN YEMEN

Trevor Hugh James Marchand

W0234769

Routledge
Taylor & Francis Group

LONDON AND NEW YORK

This edition published 2012 by Routledge

2 Park Square, Milton Park, Abingdon, Oxfordshire OX14 4RN

711 Third Avenue, New York, NY 10017

Routledge is an imprint of the Taylor & Francis Group, an informa business

© 2001 Trevor Hugh James Marchand

Typeset in Sabon by LaserScript Ltd, Mitcham, Surrey

First issued in paperback 2012

ISBN 978-0-7007-1511-8 (hbk)

ISBN 978-0415-695442 (pbk)

British Library Cataloguing in Publication Data
A catalogue record of this book is available from the British Library

Library of Congress Cataloguing in Publication Data
A catalogue record for this book has been requested

MINARET BUILDING

AND APPRENTICESHIP

IN YEMEN

CONTENTS

ACKNOWLEDGEMENTS

There are a number of agencies and funding bodies which I would like to acknowledge for their generous financial support without which this study would likely not have materialised. Firstly, I thank the Canadian International Development Agency for providing me with a CIDA Award for Canadians to carry out an independent research project on traditional architecture in Zaria, Nigeria between 1992 and 1993. It was during this time, after completing my studies in architecture at McGill University, that I fostered an anthropological approach to studying the built environment. I am also grateful to the Delta Upsilon Fraternity for providing me with the John Williamson Frederick Peacock Memorial Scholarship in 1994 in support of my Doctoral studies; to the School of Oriental & African Studies, University of London, for their very generous support in the form of a Research Student Fellowship, a One Year Language Scholarship which enabled me to study Arabic, and an Additional Award for Fieldwork; and to Senate House, University London, for a Central Research Fund Grant in support of my field studies.

There have been many teachers and university professors to whom I am indebted, but I would like to especially thank Professor Norbert Schoenauer for his stimulating lectures at the School of Architecture and his encouragement to look beyond the paradigms of Western architecture in considering space, building and aesthetics; Professor Jerome Rousseau for cultivating my interests in cognitive anthropology, and to whom I aspire as a teacher; and to Professor David Parkin who acted as my second supervisor during the early stages of my doctoral studies, and who encouraged me to cultivate a position within the philosophy of mind debates on the issues of mind-body, consciousness, and intentionality. I am also indebted to Professor Stephen Hugh-Jones and Dr. Gabriele vom Bruck who so generously took the time to read the manuscript and gave me a mine of suggestions; Dr. Jennifer Law for her always stimulating conversation and feedback; to Maeve Haldane for her wonderful editing and advice; and to Shelagh Weir for her insightful counselling on preparing the manuscript. Above all, I thank my supervisor, Professor Richard Tapper, for his advice, wisdom and support; for his ideas and experience which he shared so generously with me; for his close and critical reading of all my written work; and for being a good friend who always found the time to see me and to discuss my project and other affairs. My study is largely dedicated to the subject of apprenticeship, and it was during my 'apprenticeship' to Richard that I learned more than just how to be a scholar, but also learned a great deal about being an academic 'person'.

Of the many people I would like to thank in Yemen, I must begin by thanking the family members of the Bayt al-Maswari, and especially the head of this remarkable family of Master Builders, Muhammad al-Maswari, for accepting me as a labourer on their minaret projects, and for providing me with the experiences and information which fill these pages. I am also indebted to the incredible hospitality of Debbie Dorman and Mike Gowman who welcomed me into their tower house in the old city of Sana'a where I lived throughout my stay in Yemen. I would also like to thank 'Abdullah al-Maswari, 'Abdullah al-Hadrami, Shaif Jaralla, Walid al-Manwakil, Ahmed al-Biel, Katherine Gundel, Tim Mackintosh-Smith, 'Abdulwahhab al-Sayrafi, Dr. Hassan 'Abdulwahhab al-Shamahi, Muhammad al-Shahari, and Ursula Dreibholz for their friendship and their contributions of knowledge about this fabulous country, its history, customs and language. Thanks also to Dr. Hatim al-Sabahi and Dr. 'Abdullah Zeid Ayssa of the Sana'a University Department of Architecture for their time and support. I would also like to thank the American Institute of Yemeni Studies for providing me with an initial base to orient myself and a centre to meet a host of fascinating scholars working on various topics of research, as well as the British Council, and particularly Brendan MacSharry who made all those from the Western, African, and Yemeni communities who came to the Council-sponsored events feel genuinely welcome there.

I am eternally grateful to my friends and family who stood by me throughout my studies, and supported my academic objectives to develop an anthropological approach to studying architecture. Esteban Ayerbe, Dr. Jennifer Law, John Diodati, my brother Keith, and my sister Linda and her husband Eric, made sure that I sustained interests in various aspects of my life outside of the often all-absorbing realm of research. Lastly, I thank my parents who have encouraged all their children to pursue their passions and have served as an inspiration for the three of us. I dedicate this thesis to them, and in memory of my mother who I so wish could have seen this in print.

PREFACE

Minaret Building and Apprenticeship in Yemen focuses on a team of traditional builders specialised in the construction of towering mosque minarets in Sanaʿa, Yemen. Through a combination of vivid architectural and ethnographic description, this study of apprenticeship and human spatial cognition provides a unique insight into the daily lives and activities of a professional class of craftsmen who might be classed as 'petty bourgeois'. Although anthropological research in Yemen has addressed the tribes, fishing communities, and the *akhdam* (impure classes of sweepers and barbers), there have been no such studies on the professional craftsmen of the country, nor on the unique teaching-learning processes that distinguish their trade and mould both their professional and social characters.

Between September 1996 and August 1997 I worked as a building labourer for the Bayt al-Maswari family in order to *learn about learning* in a context in which formal technical training, engineers, and drawn plans are non-existent. The Bayt al-Maswari have specialised since the early 1980's in minaret construction and have been largely responsible for the renaissance of graceful Sanaʿani-style minarets that pierce the skyline of the newer city quarters. Throughout the fieldwork, I subjected my practical experience on site to a hermeneutic phenomenology in order to better understand the manner in which expert trade knowledge is inculcated, fostered and reproduced. The narrative of the study operates simultaneously along several different levels, pointing to the interconnection between historical, religious, architectural and social factors which shape the traditional knowledge and discourse.

On one level, the study's historical and aesthetic analyses of minarets connects it to a line of specialised twentieth-century works on minarets from Creswell (1926) to Bloom (1989), with the additional dimension of a complete technical description of their construction, and an insight into the duties of the builders and motivations of their patron-clients. Though minarets as a building type are addressed in several works on Yemeni architecture (al-Hajari, 1942; Finster, 1992; Serjeant & Lewcock, 1983;

Lewcock, 1986), none address the subject in a fully comprehensive manner. This study also seeks to compliment existing specialised studies on Yemen's domestic architecture such as Bonnenfant's (1995) and Golvin's (1984), as well as the body of more general works on Yemeni architecture (Bel, 1988; Bonnenfant, 1995; Costa & Vicario, 1977; Damlugi, 1992; Golvin, 1984; Hirschi & Hirschi, 1983; Lewcock, 1986; Varanda, 1982), by focusing fully for the first time on the builders as social agents and reproducers of an expert knowledge.

During the last two decades, there has been a considerable amount written in the anthropology of architecture (for example: Bonnenfant, 1995; Hugh-Jones & Carsten, 1995; Kent, 1987; Knapp, 1989; Melhuish, 1996; Nitschke, 1993; Waterson, 1990 etc.), as well as on the issues of space and place (Auge, 1995; Bender, 1993; Casey, 1997; Feld & Basso, 1996; Fox, 1997; Hirsch & O'Hanlon, 1995; Moore, 1986; Pandolfo, 1986 etc.). Architects and architectural historians have also adopted increasingly anthropological and sociological approaches in their analyses of built forms and urban environments (Oliver, 1987; Rendell, 1999; Rykwert, 1988). However, none of the publications seriously address the lives and roles of the actual builders, the manner in which they learn their trade, acquire a mastery over the expert discourse, conceptualise space and make on-site decisions. Likewise, other authors have not engaged with their field research as simultaneously anthropologist, architect and building labourer.

Minaret Building & Apprenticeship in Yemen addresses a significant gap in the anthropology of architecture, and seeks to move the discipline forward onto a new level of understanding. The study advances an understanding of the built environment in a direction away from static symbolic/semiotic analyses toward one of 'process', and studies architecture and urban space in its *making*, both technically and culturally, locating that understanding in the agents that produce it. I have drawn upon my experience as an architect and my training as an anthropologist to develop a detailed ethnographic account of my subject, overlaid with technical insight, and framed by analytic theory comprising cognitive anthropology and philosophy-of-mind.

The emphasis on learning and spatial cognition is integral to the work and addresses currently popular issues of the mind-body relationship, the 'modularity' thesis of spatial cognition (Fodor, 1983; Jackendoff & Landau, 1992), and human intentionality. It seeks to move beyond the notions of embodied knowledge put forward by Ryle (1949), Bourdieu (1977), Jackson (1989) and Connerton (1989), and closely investigates the manner in which this knowledge is acquired and reproduced by engaging closely with recent cognitive theory (Bloch; Dennett; Fodor; Jackendoff; McGinn; Sperber; etc.). It is hoped that the integration of these issues with an anthropological field method will make some small contribution to contemporary research on the human mind and learning.

LIST OF MAPS, PLATES AND FIGURES

Maps

Plates

Figures

All photos and drawings by the author except plates 37 and 55; and figures 17 and 18.

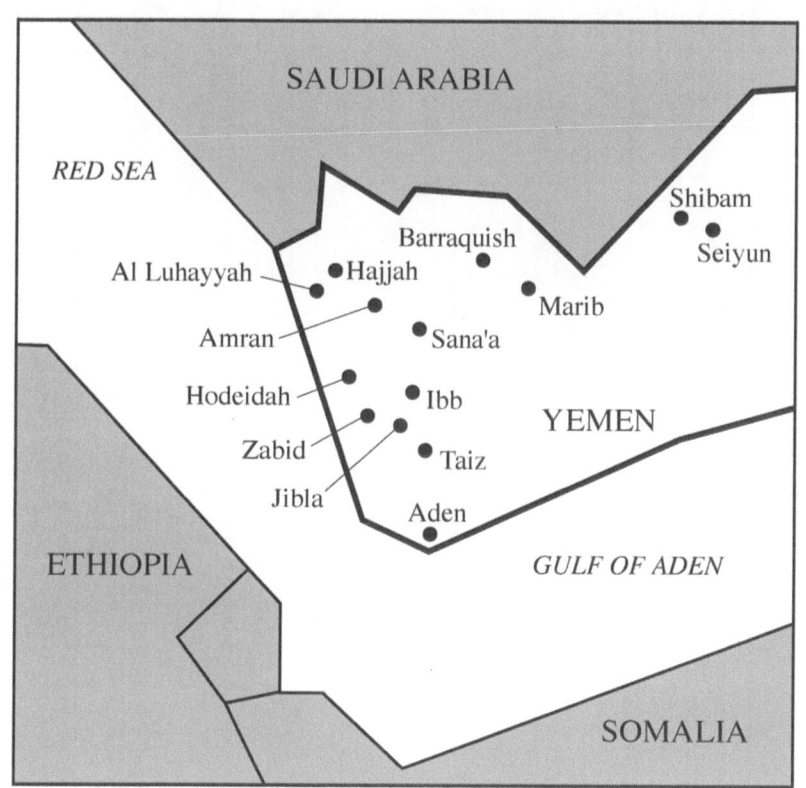

MAP 1 General map of Yemen

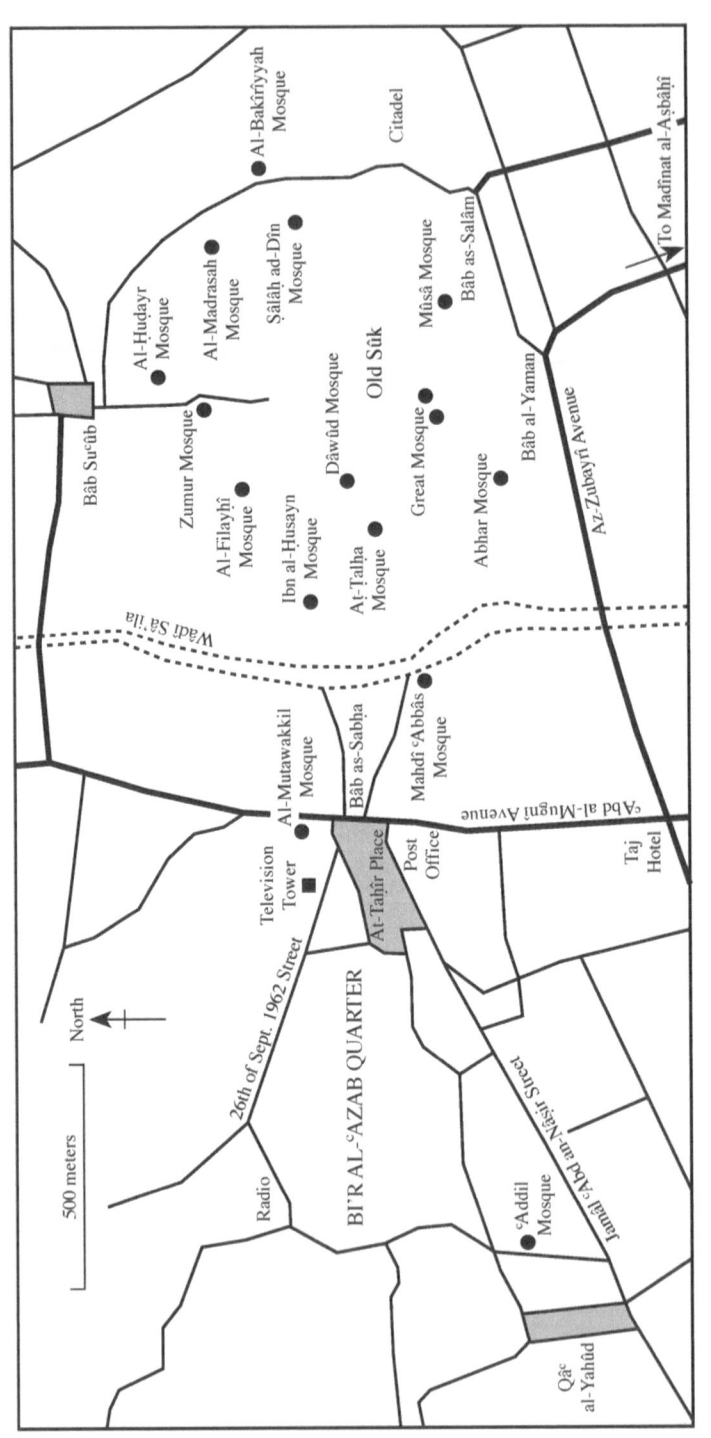

MAP 2 Map of Sanaʿa

Introduction

SANAʿA, CRAFT AND THE BUILDING TRADE

In the Spring of 1990 I paid a visit to the Zamana Gallery in Knightsbridge to view an exhibit of large and extraordinary ciba-chrome prints of Yemeni architecture taken by the Maréchauxs. A few months later, accompanied by a fellow architect and our *Lonely Planet* guide, I was peering down from an aircraft onto the mountain-cap villages of the Yemeni highlands on a journey to the capital city, Sanaʿa. My first visit left me, as an architect-turning-anthropologist, deeply impressed by its visual splendour and sophisticated sense of space. I continued research on Yemen and the country's indigenous building practices when I returned to my home city, Montreal, and later presented several lectures to my alumni at the McGill University School of Architecture. A subsequent research project in 1992–93 on vernacular architecture and extended family compounds in the Hausaland city of Zaria, Northern Nigeria, enabled me to develop an anthropological approach to my studies of space.[1] In 1996 I returned to Sanaʿa to conduct field work for a year-long period extending into 1997, and again for a shorter visit in 1998 in order to advance both my theoretical and methodological approaches to studying spatial cognition and apprenticeship amongst traditional builders. Sanaʿa al-Qadeema, or Old Sanaʿa, presented an urban fabric rich in history, built forms, and aesthetic innovations which inspired my aesthetic sensibility and intellectual curiosity. In the first section of this introduction I will outline the historic and physical context of the city before proceeding to a discussion on the contemporary condition of apprenticeship and craft in Yemen.

I Sanaʿa: An Overview of the Historic and Physical Context

Popular legend claims that Shem, son of Noah, was the original founder of the city. Whether accurate or not, Sanaʿa is reputed to be one of the oldest inhabited urban centres in the world, and it continues to play an important role as the administrative capital of a recently unified Yemen and the most populous city of South Arabia. Centrally situated on an immense highland

PLATE 1
View over the old city of
Sana'a with the minaret of the
Dawud Mosque in the centre-
ground.

plateau at an altitude of 2300 metres, Sana'a's prominence was historically established by its strategic nodal location along overland trade routes. Most important of these were the Frankincense caravan routes conveying the produce of the southern Hawdramawt region to the markets of the Nabataean Kingdom at Petra and the cultures of the Mediterranean rim. Although commodities in the *suq* have changed considerably over the centuries, and particularly over the last few decades, Sana'a still thrives as a trading centre and its prosperity continues to be reflected in the commanding architecture of the merchant tower houses and splendid minarets of Sana'a al-Qadeema.

A combination of geographical remoteness and rigid government policies (most recently those of the twentieth century Imamate) ensured restricted forms of direct contact between the Yemeni population and the rest of the world throughout various lengthy periods of the region's history. Within the country the population continues to be dispersed between a few small cities, many towns, and a vast number of villages and hamlets, which have traditionally maintained their own separate tribal identities and engaged in only limited co-operative endeavours with one another. Following the revolution in 1962, which overthrew the last of the Zaidi Imams who had ruled North Yemen since 1918, a republic was proclaimed, officially opening borders to foreign interests and initiating the construction of a road system which aimed to connect the country with its capital. Nevertheless, nearly forty years later, demographic patterns have remained relatively unchanged. Most of the country can still be classified as being in early stages of development by international standards, and, prior to the Gulf War, most rural worker migration had been to other countries rather than to Yemeni cities. As a result of sustained dispersal, the governments of the Yemen Arab Republic rarely had effective control outside the triangle formed by the cities of Sana'a, Taizz, and al-Hudaydah, measuring no more than 250 kilometres per side.[2] As Burrows comments "the history of YAR, before integration as a unified Yemen in 1990, was largely a history of new roads as agents of change and integration" amongst traditionally hostile populations.[3]

Despite its protected status of *hijrah* accorded by the seven surrounding tribes, Sana'a has been sacked on several occasions, most recently in 1948, necessitating defensive town planning, and forcing the conclusion that the concept of *hijrah* revolves more around definitions of tribal identity than on sacred attributes.[4] As an eminent trade and administrative city encompassed by territories of ideological diversity, Sana'a has likely always been fortified, though, as Serjeant suggests, not necessarily walled throughout its history.[5] The old city itself, now comprising less than ten percent of the whole city in terms of population and economic activity,[6] is divided into four quadrants, excluding the market area, and each contains a number of smaller city quarters called *harat*.[7] Extended families of common clan lineage, or clans of common tribal association, had normally settled in the same quarter which took on the clan identity, and divisions between quarters reflected inter-clan relations and hostilities.[8]

Each of the quarters traditionally had a headman, the *'aqil*, who was responsible for dealing with petty disputes amongst residents, for collecting money to finance local projects, and for maintaining the cleanliness and security of the streets in the quarter,[9] though, over the past two decades, much of their former responsibility has been absorbed by the apparatuses of an increasingly centralised city administration. Street addresses are localised by the urban geography, often referring to trade groups, individual persons, prominent geographical points, or, most commonly, to the local mosques around which the *harat* are centred. The streets are hierarchized from public to private, becoming narrower and bent as one penetrates deeper into the tighter residential realms, and opening up into small public spaces at junctions between streets or quarters.[10] Bonnenfant estimated in 1989 that approximately twenty percent of the landscape in the old city included gardens, or *bustans*, connected to mosques, and *maqshamas* (walled fields for the production of food).[11] The ground level of many of these green areas is below street level as they would have originally served as the borrow-pits from which laterite was taken for the fabrication of building bricks.

The houses of Sana'a al-Qadeema are unique, reaching six to eight stories of combined stone, kiln-baked clay brick, and occasionally mud construction, closely competing with minarets for domination of the skyline. Their profuse display of exterior decoration contrasts sharply with the generally stark exteriors of residential architectures of North Africa and most regions of the Middle East. A vast array of decorative motifs draws on abstracted floral designs, geometric patterns, and a repertoire of symbols, usually balanced in unique combinations, demonstrating both innovation and maintenance of an historical tradition: some motifs depict modern conveniences and technologies while others can be traced to ones found in the city's early mosques. Over the last two centuries, *gamariyyah* windows, elaborately carved from plaster and inlaid with coloured glass, have gradually replaced the simpler circular windows paned with thin sheets of alabaster, though several good examples of the older 'moon stone windows' survive.[12] We know from the accounts of Carsten Niebuhr, a German member of the Danish Expedition which visited Sana'a in the mid eighteenth century, that city noblemen had already begun installing Venetian coloured glass in the windows of their houses.[13] In rhythm with the white-washed decorative brick relief, an elaborate assortment of wall perforations, from main windows and fan-lights to ventilation slits, has continued to play an important role in the aesthetic composition of public facades.

Despite the seemingly overt nature of Sana'ani residential architecture, houses are in fact designed and constructed to ensure maximum privacy for the (female) inhabitants. In the manner that the first *sura* of the Qur'an is used to open the textual space of official documents in a Derridian sense, "In the Name of God the Merciful and Compassionate" is often inscribed in the lintel above the front door to the house demarcating the threshold between public and private space.[14] Street entrances to houses are positioned to avoid directly facing one another, and windows very rarely open onto the courtyards and terraces of adjacent homes. Interior functions are primarily divided in a vertical manner with each storey serving different purposes. The interconnection between them is via a wrapping stair case whose massive masonry support pillar behaves as the main structural component of the floor and roof system. The walls of the ground storey are constructed in large right-angled blocks of stone, and narrow ventilation slits lend the typical residence an impregnable air. This lower part of the building is used for the storage of foodstuffs, merchandise, and livestock, and in many of the older homes also contains a water well which may be accessed from the upper stories through a series of openings in the floors.

However, most, if not all, homes are now supplied with piped water from the city's main supply.

Upper stories are used by the resident women and children of the extended family, each wife being allocated her own private sleeping quarters and sharing common reception rooms with the rest of the female contingent. The extended family residence is nearly always patrilocal, and in a male's father's house a married couple normally has a separate room for themselves, but may take meals with the rest of the family.[15] Kitchens are located at the lower levels, normally against the less prestigious north side which suffers poor light and cooler winds. Toilets are normally situated at half-levels off the staircase, and at one time, before the installation of sanitary plumbing, a simple engineering of the floor drains divided liquid waste, which was diverted to a lime-washed channel running down the facade of the building, from solid waste which dropped straight below to a night soil room at ground level where it was dried out and later collected. The reception rooms, or *diwans*, for adult males are located on the uppermost floors which enjoy better views, and are oriented southward to capture refreshing breezes which blow from the direction of Aden. On some houses the highest storey consists of a single room known as the *mafraj*. This is the most prestigious room of the house with a lofty ceiling, large windows, and interior plaster decoration, and is used as the principal male social space for the reception of guests and the hosting of afternoon *qat* chewing sessions.[16]

Recognising the unique architectural heritage of Sanaʿa with houses, mosques, and other public buildings of considerable date, UNESCO placed the city on its world heritage list in 1984. Since that time, local heritage movements have been established and considerable effort has gone into preserving the character of the walled city while simultaneously upgrading services in an attempt to sustain a living population in the historic core. Such projects have included the installation of running water, drainage pipes, electricity, telephone lines, sealed road surfaces, and a vast renewal of the *Saʾila*, or dry river bed that runs down from the mountains and through the city. Through local initiative, many new government buildings erected after the revolution were built with the use of polychrome stone, patterned to evoke earlier twentieth century palaces and the traditional decoration of the older homes.[17] The costly manufacture of *qaḍdad*, a highly effective impermeable coating for domes, roofs, and openings, was revived in 1987 under the direction of the Bureau of Sanaʿa Heritage after having fallen out of use since 1950.[18] Like many locally available building materials, its production had suffered a serious demise with the steady infiltration of imported products and techniques, despite the fact that many of these proved to be ill-adapted to the local conditions.

Concerted efforts to revive and maintain traditional building practices and aesthetic expression have been facilitated by slow economic growth, and thus the built urban environment of walled Sanaʿa has remained, for the most part, homogeneously intact relative to other sizeable Middle Eastern cities. Many of the Sanaʿani builders continue to employ indigenous materials and methods in the construction of homes, mosques, baths and even institutional buildings. Consequently, the role of the Master Builder survives despite the influx of contractor-developers and associated imported building technology. The cartographic work of the Italian, Manzoni, during the second half of the nineteenth century has provided a valuable source of reference for determining the extent of change that has occurred in the town planning of Sanaʿa beyond the city walls since the second Turkish occupation (1849–1918).[19] As Messick notes in reference to his earlier studies in Ibb, a comparison can be made between changes in the calligraphic world and the physical articulation of space. The organic planning of Sanaʿa al-Qadeema which engineered the curvilinear movement of its inhabitants through space once mirrored the composed authorship of the traditional spiral text of legal documents. This has since been infiltrated by the coercive powers of a regulating central authority in Foucault's sense, normalising text, like buildings and town planning, within the rational structure of a grid: houses have become specialised as residences, mosques as mosques, and schools as schools, laid out in orderly fashion like the standardisation of the legal document.[20]

As the capital and largest city, Sanaʿa is an important nexus for changes in the country, a staggering number of which have occurred since the end of the revolution in 1967. An influx of foreign workers to Yemen coupled with an exodus of Yemeni male labour to Saudi Arabia, Europe, and North America throughout the twentieth century has brought Yemeni society into direct contact with a multitude of external influences.[21] The construction of a sealed road network throughout the country has connected once nearly inaccessible regions to the capital. In similar manner, telecommunication networks, and in particular the television, have served to formally connect once 'inaccessible' Yemen with the world (by 1984 at least one television was owned per family in Sanaʿa).[22] Participation in the global economy, increased per capita income, the national trade deficit, and changes in government structure, legislation, and national security policy have also been significant catalysts for recent changes in the social base of Yemeni life.

Paul Dresch maintains that under the rule of the twentieth century Zaidi Imams (1918–62), little effort was made in developing the country, and that directly following this period, between the years 1962 and 1970, most of North Yemen was engaged in a civil war between Republican forces backed by Egypt and the Saudi supported Royalists.[23] During field studies in the mid 1970s Messick observed that the younger generation of Yemenis looked to the modern Arab states for change, while the older generation continued to eye Saudi Arabia for inspiration, indicating that the divide in North Yemen also refelected age sets.[24] It was only after the resolution of internal conflicts in 1970 that development efforts accelerated, fuelled by an influx of foreign capital sent back from the emigrant labour force working in the oil states. These remittances, which quickly accustomed popular taste to a greater number of imported goods,[25] began to ebb in 1983, conjoined with a weakening Yemen riyal. Despite the announcement in 1984 that Yemen had its own oil reserves, deposits have proven to be moderate and the economy continued to weaken.[26] These factors translated into escalating prices for imported commodities and building materials, putting most out of reach for the average consumer. In addition, the delay or abandonment of development projects due to clipped financial resources has been further agitated by the most recent civil war between north and south factions of the unified country.

Dresch suggests, however, that although the lifestyles of a great many have been seriously affected by the economic downturn, most individuals and families will not have altogether lost their former positions within Yemeni society.[27] For this reason the revolution of 1962 has been conceived by some as a misnomer as many of the long standing inequalities based on birthright position within the society have remained unaltered.[28] Others argue that the emergence of a new class structure driven by capitalist interests and urban migration has resulted in more marked opposition between newly emerging economic groups than any in the country's past.[29] Historically, the main division in Yemen was between the Zaidi population of the north and the Shafi'is of the south, who have both been established in Southern Arabia since about 628 CE, or year 6 of the *Hijra*. Differences between these two religious groups have in fact been more cultural and political than rooted in any deep divide over religious doctrines.[30] Following the revolution there was a reversal of the southward movement of Zaidi influence, with a Shafi'i migration northward and into the larger urban centres. This enterprising population was quick to harness the market potential of a new capitalist economy, playing a major role in the development of the private economic sector, and consequently giving "Sana'a a more mercantile face".[31] Many of the builders involved in the vast expansion of the capital have also been *al-Yaman al-sharaf*, or of southern Yemeni origin,[32] or from the region around Dhamar, traditionally considered the dividing point between the Zaidi North and the Sunni South.

During periods of political stability between the nineteenth and the mid-twentieth centuries, the resident population of Sana'a could be safely estimated at a somewhat steady figure between thirty- and fifty-thousand.[33] Urban migration following the revolution multiplied traditional figures nearly tenfold, reaching an estimated three-hundred-thousand people by 1984,[34] and doubling to approximately six-hundred-thousand by 1990 following unification and the return of migrant workers from the Gulf region.[35] The construction of Maydan al-Tahrir Square, positioned at the nexus between the old city and the former Turkish quarter of Bir al-'Azab, and two new commercial arteries adjoining it, significantly drained the livelihood away from the old city *suq*, and physically severed the hinge between those two intersecting circular components of the walled city which had once formed a figure-eight layout.[36] The new streets were lined with brightly-painted, two-storied reinforced concrete buildings, with commercial space on the ground level and residential flats above, and signifying Sanaʿa's coming into the modern age. The subsequent completion of an outer ring road around the city in 1974 proved to be the next major step in a mushrooming growth that led to land speculation and sky-rocketing land prices.[37]

Urban expansion has consumed a great deal of the gardens and agricultural fields that once covered ninety percent of the Sanaʿa plateau: according to Saqqaf the city covered about 740 acres in 1962, and grew to 7500 acres by 1989, with construction permits for new homes increasing threefold during the peak economic years of 1973 to 1983 alone.[38] The younger generations of wealthy favoured the lower villa-type residence over the tower house, and women have also adopted preferences for new materials like washable house paints, and kitchen and bathroom equipment which require less cleaning.[39] This populace has continued to move away from the city into the expanding suburbs, where lower densities, and therefore less efficient distributions of public amenities and sanitary services, strain the city's already frail economic capacity. By default, the historic centre has become the recepient of the poorer rural immigrants who arrive without the financial means to maintain the existing urban fabric they have inherited,[40] though this trend has subsided considerably. This has also led to a geographic segregation of classes which had little precedent in the economic mix of the traditional *harat*.[41]

Perhaps the most significant social change affecting the state of the built environment and the practices of the house builder is the corrosion of the extended family structure in favour of nuclear family arrangements. Although the control of women remains firmly in the hands of their fathers, their elder brothers and their husbands, the reins over young males have loosened.[42] Many young, financially capable Sanaʿanis have chosen to set up their own independent family residences away from the constricting pressures of the older generations and the continual surveillance of the extended family. Within a short period of time, many of the once-necessary multi-storied family homes so typical of walled Sanaʿa were abandoned for newer single-storey bungalows outside the city centre, or were subdivided into smaller self-contained flats. Like other regions of the Muslim Middle East which have experieced this 'modernising' phenomenon of suburban expansion, the women disengaged from the traditional environment of kin and neighbourhood networks found themselves increasing secluded and their roles ambiguously defined in the new economic order.[43] However, despite the persistence of this trend, in more recent times it has once again become fashionable among certain wealthier segments of the old-city population to remain in their traditional tower houses. In the latter half of the 1990s a number of new tower houses were in fact erected in both the old city and in the Bir al-'Azab quarter, and several existing ones were extended vertically with additional storeys.

Unlike neighbouring Saudi Arabia or Iran, Yemen is not rich in the natural resources which spawn quick development, and consequently its physical environment has not been subjected to the same magnitude of rapid change experienced by some of its neighbours. The discovery of modest petroleum deposits in the eastern part of the country in 1984 also coincided, fortunately or unfortunately for Yemen, with the drop in world oil prices, hampering any quick and extensive development ambitions. Also unique in Yemen's development policy is the introduction of Local Development Associations, or LDA's, which have been set up through local initiatives to sponsor and administer small scale development projects in direct response to regional needs.[44] Carapico points to such organisations as being one of the factors which have given rise to civil society and democratic progress in Yemen.[45] It is hoped that continued efforts on the part of local development schemes will demonstrate a sensitivity to the existing infrastructure, and work to preserve it rather than efface it. In far too many instances the historic urban fabric and traditional construction techniques and materials have been disregarded by consultants who have implemented costly modernising programmes which are inconsequential to the actual needs of Yemenis, and incommensurable with the country's financial capacity.

II The State of Apprenticeship and Craft in Yemen

The crafts and sciences are the result of man's ability to think, through which he is distinguished from the animals ... [The sciences and the crafts] come after the necessities. The [susceptibility] of the crafts to refinement, and the quality of [the purposes] they are supposed to serve in view of the demands made by luxury and wealth, then correspond to the civilisation of a given country.[46]

Ibn Khaldun

Ibn Khaldun's fourteenth-century association between the crafts and the sciences is echoed in the contemporary writings of both Titus Burckhardt and Seyyed Hossein Nasr who stress the critical link between the two in the Islamic world. Nasr goes so far as to assert that Muslims "never distinguish between technology and the crafts, or between the crafts and art",[47] and Burckhardt suggests that any split between them is "a relatively recent European phenomenon". The latter continues that:

9

formerly, every artist who produced an object was called a craftsman, and every discipline which demanded not only theoretical knowledge but also practical ability was an art ... Art always involves technique (*san'ah*) and science (*'ilm*).[48]

The requisite techniques and sciences of an art were transmitted from one generation of craftsmen to another through a form of apprenticeship, often affiliated with guilds, in which individuals were trained by a Master craftsman in their home or atelier. Parallels have been drawn between the typical relations of the guild master and his apprentices with those of the Sufi Master and his noviciates, and the impact of religious education on the formation of a Muslim craftsman has been widely noted. An Islamic emphasis upon God's Omnipotence inhibits the occurrence of "any kind of Promethean humanism", and religious training gives rise to "the inculcation of certain attitudes and the elimination of other possibilities" which shape the minds and practices of the artisan.[49] Eickelman observed that, in Morocco, the methods by which craft knowledge is conveyed are analogous with the transmission of the religious sciences, and both engage a "quasi-genealogical chain of authority which descends from Master or teacher to student".[50] It would not be reasonable to assume that all religious noviciates reach the peaks of spiritual enlightenment, nor that every apprentice masters his or her trade, but in each case it can be expected that some degree of expert knowledge does get passed along and certain levels of understanding are achieved. To quote from Nasr:

> Not all those who learned to weave carpets or make tiles were consciously aware of the profound metaphysical and cosmological significance of the symbolism of patterns, forms and colours with which they were dealing, nevertheless, something of the science involved was transmitted from the beginning, the knowledge becoming more explicitly elucidated as the student advanced in the mastery of his craft and came to gain a more immediate awareness of the nature of the materials with which he was working and the principles by which his art was ennobling the material he was moulding.[51]

However, these idealised conceptions of a unity between the arts, crafts and sciences in the Islamic context, and of an interface between religious systems and the inculcation of trade-related ideas and performance in the apprentice have been challenged by other writers as being no longer relevant in the contemporary world. Yusuf Ibish, writing generally on the role of the guilds in Islamic Middle Eastern cities, identifies the nineteenth century as the beginning of the end for many apprenticed crafts in the Muslim world. Their decline has been propagated by a number of complex historical factors rooted in the processes of globalisation that burgeoned with European colonialist ambitions. Ibish points out that as a consequence of expanding foreign interests and capitalist ventures there was a rapid displacement of many locally made commodities by an influx of mass produced European goods.[52] Also, the reorganisation of government structures in accordance with colonial, and later Western models, and the subsequent introduction of new taxation policies, served to weaken the power and financial base of many craft guilds. In agreement with this, Lewcock notes that since the time of the second Ottoman Turkish occupation in Sana'a municipal organisations have largely eclipsed the responsibilities and status of the craft shaykhs and shaykhs of the market.[53] Throughout the region, modern transportation systems altered traditional trade routes leaving once important trade centres obsolete, and population explosions over the past two centuries have resulted in mass urban migration to Middle Eastern cities which continues to deteriorate the delicate symbiosis of social and economic ties of historically established communities. Finally, but not conclusively, Ibish states that the adoption of Western models of standardised secular education has had a weighty impact on both the consumer tastes and the career choices made by the younger generations.[54]

Indeed, during the decades between North Yemen's revolution and the Gulf War, the prospects for many of the country's handicraft industries appeared to reflect Ibish's ominous assessment. This period was marked by an enormous emmigration of young Yemeni males to the neighbouring oil-rich states for work and the influx of imported goods, ideas and lifestyles. Remittances sent back by emmigrant labourers fueled a quickly growing economy and consequently, a booming modernised construction industry which offered higher salaries to prospective young builders in comparison to the meagre wages payed by the local traditional Master Builders. In reports issued by both the Italian organisation, Studio Quaroni,[55] and Ronald Lewcock[56] during the 1980s, strong recommendations were made for the establishment of institutions and schools which might ensure the survival of traditional crafts (including those of the building trade) against the onslaught of imported mass-produced goods and technologies. Some handicrafts, however, continued to prosper during this period. In the early 1980s, Walter Dostal found that among the market crafts which were still practised primarily in Yemen's northern urban centres, the production of arms had monopoly in terms of the vitality of the trade. This was still much in evidence during the time of my own research more than ten years later by the prolific production of *jambiyyah*s – a curved dagger worn by the majority of men in the north of the country, and which comprises the separate manufacture of the blade, handle, sheath and belt.[57]

The effects of the country's general economic decline following the Gulf War and mass return of migrant labourers have restored market conditions which support some forms of local craft production. Shelagh Weir has noted that in addition to the revival of some traditional Yemeni handicrafts, there has also been an emergence of new craft industries over the past three decades in response to the availability of new production materials and technologies, and changing consumer demands. These have included, for instance, the fabrication of decorated and coloured metal doors which have largely replaced their wooden counterparts in most new constructions, and the production of various rubber wares from recycled tires (which include the stitched rubber buckets used to transport supplies and materials in the building industry). Based on her recent participation in a survey of handicrafts in Yemen, Weir considered that the state of artisanal production in many of the crafts was reasonably healthy, and did not necessitate significant intervention by the government or by foreign development and conservationist interest groups.[58] This supports my own findings that many of the apprenticed crafts, and in particular those affiliated with the building trade, were still operational in Yemen and continued to play a critical role in the economy. In addition to families (*bayts*) of builders who were specialised in the construction of such building types as houses, public baths, mosques, and minarets, other craftsmen were also actively training apprentices to perpetuate their respective expertise including carpenters, decorative plaster workers, *qaddad* (lime plaster) workers, and decorative coloured-glass window (*gamariyyah*) makers (see plates 2 and 3).[59]

In the remainder of this section, I will consider several of the important factors which have thwarted the complete demise of Yemeni crafts and their associated apprenticeships, as well as those that have instigated those changes such as those identified by Ibish. Briefly, these include a history of minimal foreign intervention relative to the histories of other regions of the Middle East; the foreign impact on the formation and development of a state education system; the history of trade routes through Yemen, and the status of imported goods in the market place; population growth and fluctuations in the current economy; and the recent processes of democratisation. I will briefly argue that the final issue constitutes perhaps the most formidable threat to the master-apprentice relationship and the unique teaching-learning processes that characterises it.

Foreign Intervention in Yemen

Historically, North Yemen, and notably the highlands, was never successfully conquered and maintained by a foreign power, and until the Revolution of 1962 the region remained relatively isolated from the mainstream of global politics. It is true that a plenitude of foreign rulers filled the vacuum left by the Himyars, last of the great South Arabian Kingdoms: the Axumites established a seat of rule in Sana'a after their third

PLATE 2
A young apprentice chiselling
out the form of a *gamariyyah*
window in finer, smoother
detail.

march on the country, and the Sassanian Persians stayed on as rulers
following their role in the expulsion of the Axumites in the late 6th century
AD, but foreign rule was constantly challenged and distance from the
homelands served to complicate affairs. After the advent of Islam and the
death of the Prophet, the *rashidun* Caliphs extended their dominions over
the country, and the ensuing Umayyad Caliphate, followed by the
'Abbasids, and so on, reigned precariously over this distant land through
often self-serving governors who spun their own dynasties on Yemeni soil
and wove their glories into the tapestries of a native history: the 'Abbasid
Governor Muhammad ibn Ziyad severed ties with Baghdad and founded
the Bani Ziyad Dynasty; Ali ibn Muhammad as-Sulayhi, representative of
the Fatimid Caliphs of Cairo, conquered the Ziyadids and, in turn, founded
the Sulayhid Dynasty; and Umar ibn Rassul, also known as Turanshah, was
sent as a vizier to the Ayyubid Sultan Salah ad-Din, but in due course
assumed the title of al-Mansur ('The Victorious') and established the
Rasulid Dynasty which flowered into one of the greatest empires in the

13

PLATE 3
A Master Craftsman displaying
a finished *gamariyyah*
window with various colours
of inset stained glass.

history of Islamic South Arabia. Even Yahya b. al-Husayn al-Hadi ila
'l-Haqq (d. 911), founder of the Zaydi sect of Shi'i Islam in Yemen and an
Imamate that survived over one thousand years until 1962, was also an
immigrant whose accomplishments and legacy became part of the regional
history and glory.[60]

Only the Ottomans were able to establish any significant periods of
foreign rule over the country, first during their occupation from 1538 until
1630, and again in 1849 to 1918,[61] ending with the final collapse of their
Empire. However, Ottoman control over Sana'a and the northern highlands
was always tenuous at best, and officials found themselves netted in
continuous processes of negotiation with the tribes. South Yemen remained
a British protectorate until 1967, but the North resisted the Western
colonial powers, and reinstated home rule under the political and religious
yoke of the Zaydi Imams following the expulsion of the Turks. The
twentieth-century Imams, Yahya and his genealogical successor Ahmad,
maintained policies of considerable isolation over their domains. Only after

the overthrow of the Imamate and the establishment of a republican government was this pocket of South Arabia fully exposed to foreign political ideologies (notably democratic ones since the period of unification of North and South Yemen in 1990), world market forces, and models of state education.

State Education

State education, which was launched in the late 1960s following the civil war and expanded in the 1970s with financial assistance and expertise from the Egyptian government, has been heavily modelled on foreign systems, most significantly on Egyptian secular education.[62] At that time, religious instruction was limited to one hour per week in the Yemen Arab Republic.[63] The school system has remained dominantly secular since the integration of the two formerly distinct systems of the North and South, but the central government in Sana'a has progressively bolstered the religious content of the curriculum. This move has been largely a strategic political response to the pressures from the main opposition Reform Party, or Islah, as well as to appease other potentially hostile religious conservatives in the country.[64] Where the government lost ground was in the struggle to keep the battered educational system effectively operational when confronted by a high average of eleven percent yearly increase in absolute enrolment figures.[65] The deteriorating quality of all levels of education was further exacerbated by the cessation of Arab aid following the Gulf crisis which forced the financially impaired Yemeni government to terminate the contracts of some eighteen thousand foreign teachers, and to fill the vacuum largely with under-qualified nationals.[66]

Despite the increase in actual student numbers, the percentage of children ages six to fifteen enrolled in basic education (i.e. grades one to nine) has remained low at about fifty percent, of which both the ratios of boys to girls, and the percentage of urban to rural children attending school are nearly 2:1.[67] The percentage of those continuing education at the secondary level thinned out drastically, especially amongst the rural population, and significantly fewer continued on to university. The university system, unquestionably the most expensive level to operate, sorely lacked the funding it once received from Kuwait, and the quality of education has continued to degenerate since the Gulf War. There were exceedingly few employment opportunities for university graduates and their Yemeni diplomas carried little prestige or promise outside the country. The elite careers that were once attainable for enterprising young scholars no longer existed. Although many young people harboured aspirations for a life of Western comforts, there were in the vast majority of cases scant means for procuring such a future.

Trade Routes and World Markets

While there is perhaps truth in Ibish's assertion that standardised secular education has had an impact on consumer tastes, the actual numbers of state-educated consumers in Yemen has remained low with a national literacy rate of less than fifty percent.[68] Therefore the education factor does not provide a sufficient explanation for the quantity, variety and appeal of imported and mass manufactured goods that have been made available on the markets in the last three decades. The availability of these types of consumer items has not been restricted to the larger urban centres where the more educated segments of the population are concentrated, but many can also be found in rural *suq* throughout the country. Nevertheless it should be noted that imported goods, as well as handicrafts, predominate in the urban markets in contrast to the greater emphasis on domestic products in the rural weekly markets.[69] I would suggest that, even with regard to Sanaʿa, the availability of imported goods in Yemen cannot be attributed to a colonialist legacy since, as I have discussed above, North Yemen's history of foreign domination is notably different from other countries of the Middle East, nor does it correlate directly with levels of formal education.

The history of Sanaʿa is largely the history of the merchant-traders who, as early as the period of the pre-Islamic South Arabian Kingdoms, settled on the high plateau at the strategic crossroads of the trade routes coming from Marib in the East to the Red Sea, and from Aden in the south to the Mediterranean. From early times, perhaps as early as the first few centuries AD, Sanaʿa had special status granted by most of the northern tribes as a *mahram*, inviolable place, or *hijrah*, protected enclave, sanctioning peaceful commerce amongst traders and buyers of different, and possibly opposing, tribal affiliations.[70] Third-century inscriptions testify that Sanaʿa was a *haram* district, and served as a military center for the Sabaeans.[71] Beeston hypothesised that the name Sanaʿa itself means 'well fortified' "since in the Sabaic language derivatives of the root sn^c are (with a solitary exception where the meaning is uncertain) exclusively associated with military defensibility", although there is still no consensus on this definition.[72]

Historically, Yemen had been a key intermediary player in the trade networks connecting the Indian Ocean with the Mediterranean. For the most part, Yemen has been a place where things *came through*, and not where things *came from*. The bulk of goods that passed through the hands of Yemeni merchant-traders were from India, China, and the East Coast of Africa. Exceptions included the aromatic resins such as frankincense and myrrh which were cultivated mainly along the Arabian Sea Coast in Dhofar (now Oman) and Mukalla, and on the Island of Socotra, and which were in enormous demand by the Mediterranean cultures in the centuries leading up to, and for some time following, the Christian Era. It is also known from records that Yemeni-produced silks and other fabrics, as well as (ambergris,[73] wars-dye [yellow], mules and donkeys) were also exported in Medieval times.[74] After a lengthy interval following the demise of the Incense Roads, coffee (again a harvested produce and not a craft) was the next significant Yemeni commodity to merit world market attention. Its introduction to foreign sailors in the Red Sea at some point in the fifteenth century soon attracted European trading fleets, and the subsequent establishment of Dutch and British 'factories' in the port of al-Mocha in the early seventeenth century.

Despite the level of activity seen by the coast, the interior has experienced only brief periods of foreign intervention, and, in fact, when Imperial ambitions turned toward the region in the early nineteenth century, it was Aden, not al-Mocha, that Britain made its object of desire. Aside from a more favourable climate and superior harbour, Aden was better positioned geographically as a coaling station on route to India, especially once the Suez was opened in 1869. For the most part it was the grand emporiums of Cairo and India at either end of the trade route, and not local market conditions and products, that steered foreign interests in the region. It is nevertheless curious that despite a remarkably long history of urban settlement and contact with an international gamut of traders and producers, the overall standard of craft production in Yemen remained very low in comparison to other countries with which Yemen had contact through the trade networks: notably India, Egypt, Syria, Iraq and Iran. During his travels in the first quarter of the nineteenth century, Robert Finlay noted that "Of the manufactures and commerce of Sana'a there was little to be said. There were the traditional gold and silver workers, mostly Jews but some Banians. Only the Jews were allowed to mint coins. Some coarse woolen carpeting was made as were the ubiquitous jambiyyahs".[75] The few significant exceptions of mainly Jewish work, such as the fine silver craftsmanship, have largely disappeared since the mass emigration of that population in 1949. In fact, Dostal records that at the time of his survey in the early 1980s only six of the twenty-seven remaining Muslim smiths in the silver *suq* of Sana'a had ever apprenticed with a Jewish craftsman, and those for only a very brief period before the community's rapid departure to Israel via the British-sponsored Operation Magic Carpet.[76]

In the last few decades, provisions and household goods sold in the Yemeni *suqs* have been dominated by imported goods of mainly Asian origin. Imaginative displays of brightly coloured plastic wares, rubber merchandise, and clashing patterned polyester fabrics battle for attention in every shop front. The type of manufactured goods may be new, but Asian wares are by no means a recent introduction to the South Arabian markets. In his discussion on the impact of imported goods on Sana'ani artisans, Mermier notes that despite claims that Yemen only entered the world market system after the Revolution of 1962, the replacement of artisanal goods by imported ones was in fact already noted by European travellers in the early twentieth century.[77] Serjeant discusses the history of the flow of trade from China to Yemen which was known to have begun before the advent of Islam and continued strong during the Medieval period. In reference to the accounts of Ibn Batutah, Serjeant writes "Sin-Kalan (Guangzou) has a large bazaar from which porcelain is exported to all parts of China, India and the Yemen", and "Cheng Ho, Grand Eunuch of the Ming Emperor between 1405 and 1433, visited Aden on three occasions, Zafar and other places".[78] The change therefore is not the sudden appearance of Asian wares in South Arabia but, rather, the sheer magnitude of the imported goods resulting from recent technologies of mass production. This continues to threaten the existence of mainly indigenous producers of household wares, fabric production, and the local smithing industry who cannot afford to make the necessary changes to their production processes. For instance, the specialised tools used by local craftsmen, such as those affiliated with the building trades, are now almost exclusively manufactured in China.

Urban Expansion and Traditional Building

Rapid urban expansion throughout the country from the early 1970s onwards, with annual population growth rates of up to ten percent quoted by some sources,[79] led to speculations that the traditional building crafts would be eradicated by an onslaught of new building materials and technologies.[80] In response, the General Conference of UNESCO at its twenty first session in 1980 adopted a resolution to work out a plan of action for the preservation and restoration of Sana'a, and the city was later added to the roster of World Heritage Cities.[81] The scope of the project has been restricted to the Old City, and has played a very significant role in bolstering both the large-scale practice of traditional construction methods and a public appreciation of indigenous materials.[82] However, in terms of land size, the Old City comprises only a small fraction of present-day Sana'a, and it can only be hoped that the urban renewal of the historic section will have an impact on the design and management of the ever-growing sprawl that fingers its way through the valley.

In 1975 the newer districts of Sana'a housed two-and-a-half times the population of the historic quarter and were seven-and-a-half times greater in area, and by 1981 these figures had become four times and fourteen times respectively.[83] Between the early 1970s and 1990, the urban population in Yemen had increased from about ten percent of the national population to over twenty percent.[84] In addition to the general trend of urban migration, another significant factor contributing to the frenzied urban growth of the decades of the 1970s and 80s was the substantial remittances sent home by migrant labourers in Saudi Arabia and the Gulf, much of which was invested in building. After the out-break of the Gulf War in 1990, eight hundred thousand migrant labourers returned home, further swelling the urban count. The massive numbers of migrants to the urban centres in recent decades has rendered any attempts to assimilate the new populations into a 'city way of life' extremely difficult. The sheer quantity of settlers empowers this group to resist conformity to the established canons. Meanwhile, the gulf of misunderstanding between original city dweller and newcomer has widened.

The discovery of oil reserves in the early 1980s, and the first well going into production in 1986, originally signalled an acceleration to the changes already sweeping the country. However, the massive loss of remittances from the returning migrant labourers after 1990, coupled with substantial short term fiscal pressures and the burden of high foreign debt, has meant a considerable slowdown in development, modernisation, and consumer spending. Political scientist Iris Glosemeyer suggested that the downturn in the economy and the resultantly high levels of corruption and patronage amongst business leaders and government, factors which have impeded the rate of Yemen's democratic process, have in fact been beneficial for modifying the rate of change in the country and salvaging traditional institutions.[85] From the point of view of the traditional building trade this has been a positive turn of events. With less money available for imported building materials and more costly modern building technologies, more clients are commissioning local builders and resorting to locally quarried stone and kiln-baked brick to dress their facades. Established families that were at one time leaving the old city of Sana'a for newer, more convenient residences in the expanding suburbs, are now reconsidering their options and many have remained and made necessary amendments and additions to their existing family houses. It was plain from the vantage point on the roof tops in Old Sana'a that the city of tower houses continues to extend vertically, and entirely new ones are being erected from the ground up.[86]

But economics were not the only factor fostering the re-appropriation of indigenous knowledges and crafts in this field of trade. Tourism, albeit a double-edged sword in the way it impacts a culture and economy, has also played a vital role in heightening a nation-wide awareness of the country's unique architectural heritage. Qadi Isma'il al-Akwa, president of the Organisation for Antiquities and Libraries, was recorded as saying:

> Until recently, we Yemenis did not have much appreciation for our architectural heritage. The people of Sana'a and indeed the people in many Islamic countries were not sufficiently aware of the treasures contained in their cities that were worth preserving ... Foreign architects and experts registered their concern over our heritage. They praised the beauty of our architecture, the types of buildings and the traditional buildings of Sana'a.[87]

A growing world-wide interest in Yemeni history, archaeology, architecture, and the country's varied natural landscape, has forced the population to re-evaluate their past in terms of both its potential as an economic resource, as well as its integral role in constituting their national identity. This latter point is a salient one, especially as Yemenis become more fully exposed to, and juxtaposed against, the multiplicity of other world cultures. There was an exceptionally strong, possibly universal consensus amongst the many Sana'anis that I spoke with, that any new building in Sana'a cannot be considered acceptable unless it is dressed in stone or brick with geometric patternings, and most importantly it must be fitted with *gamariyyah* windows (decoratively carved plaster fan lights with coloured glass).[88] Even the new American-style Pizza Hut Restaurant located on the edge of town was barred from opening by the authorities until it had installed a band of *gamariyyah* fan lights along the top of its plate-glass windows.

Architect Hatim al-Sabahi[89] championed the necessity for such architecture codes, even if they served only to address the cosmetic concerns of building aesthetics, and despite the fact that there has been no formalised chanel for enforcing such regulations to date. The buildings of Sana'a al-Qadeema were erected and decorated according to what al-Sabahi has called a "common-sense code" (*qanuun al-murasalah 'ala al-bidaya'*). This developed over centuries in response to the physical limits and potentials of the building materials, the practices and decisions of the Master Builders, and the practical needs and aesthetic tastes of the (relatively homogeneous) population. The more recent throngs of settlers from various regions, on the other hand, have arrived in Sana'a without an understanding or appreciation of the existing urban fabric and, when permitted, have imposed their own building practices and conflicting aesthetic judgements upon the cityscape. There was a notion among many native Sana'anis that enforced codes, addressing in particular the use of *gamariyyah* windows and interior decorative plasterwork, were the only guarantee of safeguarding their heritage against the ill-informed decisions of present and future builders. One close acquaintance, whose family has lived in the old city for many generations, compared these codes favourably with the disciplinary measures taken by a father in teaching his children to pray: "laws must be laid down by those who know better for the sake of the generations to come". Members of the architecture community, like al-Sabahi, have recognised however that the over-production of *gamariyyah* widows and plaster works has led to a diminished quality in these trades, and that their over-use has resulted in a decline in their once associated social prestige.

Democratic Development and Traditional Knowledge

An important factor emerging from this discussion is that there has been an increased general interest in, and the evolution of a public discourse around, the topic of architecture and building practices. The discourse has drawn competing views and interest groups, both native and foreign, into the public forum, whether it be at the university or in the *mafraj*, to discuss issues of control over public space and representation, heritage, and civic identity. The nature of such "arenas of conflicting interest" is the basic "constituting feature of civil society",[90] for "civil society encompasses not only voluntary associations but social movements and popular struggles as well"[91] responding to concrete material circumstances, and "embedded in wider struggles over the allocation of resources in society".[92] As Carapico concludes in her study of civil society in South Arabia, "among the Arab countries [Yemen] offers [the] greater hope for democratisation in the medium term", and this is largely due to "cumulative civic pressures" and not to foreign influence or enlightened governmental leadership.[93] The impact of this discourse and civic pressure on the building trade, the production of its various related crafts, and their associated apprenticeship-style training systems, is two-fold. Firstly, the public, now with a certain measure of power to choose between a multiplicity of designs and possible materials, has played an increasingly determinate role in the evolution of consumer tastes which may or may not continue to lend patronage to all or some of the existing traditional building materials and methods. Secondly, and perhaps more crucially, the public has gained increased access to a formerly restrictive and tightly controlled expert discourse.

Briefly, I am suggesting that democracy, and more particularly an eventual democratic access to knowledge, will serve to empower the general population, including the younger generation of labourers and apprentices, to question and challenge the once sacred realms of the craftsman's expert discourse. Conceivably, this would result in the break-down of traditional attitudes toward authority, such as that of the Master Builder. However, one crucial factor which has impeded this seemingly inevitable process has been the nature of Yemen's national educational system. Teachers have acted primarily as authority figures rather than in perhaps the contemporary idealised role of Western-style educators, and the school has functioned as a disciplinary establishment rather than instilling young minds with an aptitude for critical judgement and responsibility: qualities deemed to be necessary by most liberal Western political theorists for participation in civic processes and the democratic state. Carapico notes that despite the emergence of civil society in Yemen, "educational standards have never been conducive to bourgeois liberalism".[94] Classroom teachers, like the Master Builder on site, were themselves subject to authoritarian-style training processes, and therefore have been principal agents in its reproduction. Nevertheless, changes in perspectives toward authority, and increased opportunity to investigate, discuss and challenge traditionally guarded realms of knowledge will eventually lead to the de-mystification of the Master Builder's expertise in the minds of the public. Revelations about 'how he knows' and 'what he knows' (perhaps, ironically, like those discussed in the coming chapters of this book) will undermine his real power and consequently threaten the existence of the distinctive teaching-learning method supplied by the traditional apprenticeship system. This does not necessarily equate to the eradication of the building crafts, but the very nature of the trade knowledge, in terms of how it is passed along and how it is understood, will be significantly altered.

III The Scope of This Study

Briefly, the intent of my study is to develop an understanding of human spatial cognition – in other words, how do we perceive space, manipulate our conceptualisations of it, and physically act upon it. In order to do so I worked closely with an accomplished team of minaret builders in Sana'a, Yemen, modelling the basic method on my former field studies conducted with traditional builders in Zaria, Northern Nigeria.[95] In the company of the Yemeni builders, my study of spatial cognition became structured around the apprenticeship system which serves to promote a labourer to the expert status of a Master Builder via a lengthy and rigorous training. More specifically my research has focussed on the teaching-learning processes at the various stages of the training, including those that inculcate disciplined conduct and practice, an understanding of building techniques, and a mastery over spatial conceptualisation and design.

The study is presented in four chapters. The first of these aims to contextualise the mosque minaret as a building form in the history and urban context of Sana'a. With reference to several key lines in an eighteenth-century poem dedicated to the Mosques of Sana'a, and in particular to the 'Addil Mosque where I participated for the first time as a labourer on a minaret project, I reconsider the role of the tower minaret in Zaydi Yemen. I explore the relation of verticality in both Yemeni secular and religious architecture with expressions of power and piety, and investigate the sources and inconsistencies of the Zaydi's historical aversion to the minaret structure. This socio-historical study makes reference to dominant architectural theories on minarets, namely that of Jonathan Bloom published in 1989, before turning to a more specific discussion of the relation between historically significant minarets of Sana'a and contemporary designs and construction. Materials, structure, and external ornamentation will be considered before concluding the first chapter with an analysis of the minaret's physical position in relation to the mosque and the importance of its prominent vertical axis. During construction, this axis poetically mirrored the hierarchy amongst the building team who were strung out along the tower's internal winding staircase, manning their assigned work stations.

Chapters two to four are each divided into two main sections. The first and shorter section provides a technical description of the various stages involved in the erection of a minaret, and the second constitutes an analysis of the teaching-learning processes involved at the different stages in the training of a traditional builder. Both sections in all three chapters have been arranged to simulate both the chronological progression of a minaret project and the progression of a builder's career and his acquisition of an expert knowledge. I have metaphorically linked the succession of the building phases with the advancement of the craftsman's training: building the structure's foundation has been affiliated with the inculcation of basic discipline in the labourer's simple practices and performance; erecting the brick tower above the stone base and reaching the height of the calling platform has been compared with the apprentice's reaching a plane of understanding through the processes of 'making'; and completing the dome and installing the *hilal*, or crescent moon has been likened to the builder's achievement of mastery over his trade performance and intentionality, and his ultimate recognition as an *usta*, or Master Builder.

While labouring with the minaret builders I noted a correlation between the hierarchical stages of becoming proficient in the trade and the three recognised phases in achieving Islamic spiritual enlightenment. In chapters two to four, the experiences of the labourers, the apprentices, and the master craftsmen I worked with are compared sequentially with the principal stages of *islam, iman,* and *ihsan,* and relationships are also drawn with both secular and religious instruction in Yemen. Briefly, *islam,* or submission, emphasises the embodiment of disciplined religious practice; the stage of *iman* is characterised by the cultivation of a spiritual faith and the accompanying understanding of one's ritualised practices; and *ihsan* denotes the paramount level at which one's intentionality is fully absorbed in, and guided by, spiritual faith and understanding. These are compared with the bodily discipline inculcated in the labourer, the understanding acquired by the apprentice, and the intentionality which pervades the thoughts and action of the Master Builder and which qualifies his expert status.

Finally, in considering the three different positions of builders and their respective training, I develop theoretical analyses of the teaching-learning processes rooted in the ethnographic material and building upon related philosophies of mind. In chapter two, an inquiry into the inculcation of disciplined bodily practices in the labourer is inspired by the work of Gilbert Ryle[96] which forcefully challenged Cartesian dichotomies between body and mind, and which has been furthered in such contemporary works as those authored by Boyer, Connerton, and Jackson.[97] The enduring legacy of the split between body and mind has been problematic on a number of accounts, but most relevant to queries in the field of anthropology is that it deals unsatisfactorily with the causal and dialectical relation between human action and mental states. Dualism fosters the metaphysical problem of how two distinctly different substances (the physical and the mental) may interact. Traditionally, the metaphysical nature of this problem has elicited equally metaphysical responses, as might be expected. Counter-approaches to the Cartesian metaphysical tradition, early examples of which can be found in the work of William James (1892) and more importantly Gilbert Ryle (1949), have sought to undermine the dualist nature of the question, and to reformulate the problem in physicalist terms. I support this approach by demonstrating that the building labourer's training in disciplined comportment teaches him *how to act*, but not necessarily *how to understand*. In other words, knowledge is sedimented in his bodily performance, but this type of know-how does not necessitate a consciously aware form of reflection.

Chapter three concentrates on the political relations and the training of apprentices who have been chosen to work alongside the Master Builders. Following a general discussion of the instruction, responsibilities, and relationships between the young men at this stage in the training process, I address the issue of there being no distinct lexical category used by the local population in Sanaʿa to designate either 'apprentice' or 'apprentice-ship'. I proceed to demonstrate that this ambiguity of title and status is reflected in the fluidity of both the hierarchy and the division of labour amongst the workers at this level. By pursuing an investigation into the culturally determined salience of certain taxonomies and lexical categories, I introduce a theory of cognition developed by Jerry Fodor known as modularity-of-mind[98] in order to challenge and elaborate upon a Sapir-Whorf based theory of language relativity. Very briefly, modularity is functionalist theory of cognition which proposes that the mind is divided into distinct faculties which process distinct types of information and therefore produce distinct types of knowledge. These unreflective knowl-edges interact at a higher level of cognition to formulate meta-representa-tions about which we, as humans, are consciously aware and consciously act upon. Cognitive linguists employing the modularity theory claim that an analysis of language can provide a window onto the divisions of labour between the mind's faculties. Jackendoff and Landau's important study on language and spatial cognition convincingly suggests that the marked difference in a language's capacity for producing object descriptions, or the *what of an object*, as opposed to describing the location of an object in relation to other objects, or the *where of an object*, reflects an innate bifurcation in our human capacity for spatial representation.[99] After a review of this related research, I expand the currently language-biased window onto cognition by demonstrating how a careful consideration of the minaret builder's skill-related performances reveals various distinct cognitive operations. Unlike the linguistic-based theory which assesses the quantity and attributes of available vocabulary in a language for describing objects versus that of prepositions for communicating spatial relations, I have explored the processes of 'making' as a credible window onto human spatial cognition. Next, by drawing upon comparisons with my own Western training as an architect and the training of students in the Department of Architecture at Sanaʿa University, and the training of a carpenter from my native province of Quebec, I conclude this chapter by demonstrating how different forms of training – i.e. the building apprentice's mastery of the 'object' through making versus the architect's mastery of 'space' through design – result in a strong reliance upon different mental faculties for their respective spatially oriented tasks.

The fourth and final chapter is dedicated to a study of the Master Builders, and begins by introducing the senior members of the al-Maswari family for whom I worked. The methods and politics of transferring trade knowledge, expertise, and power from one generation of Master Builders to the next are considered before turning to an investigation of the minaret builder's conceptual processes of design and construction. The constitution of the craftsman's 'traditional' knowledge is explored here despite his incapacity to explain verbally 'how' he knows, and the seemingly deliberate wall of secrecy which is erected between the *usta* and his public. The inability to explain 'how' has been related to the form of training a builder receives, and I develop this analysis to further support claims made in the previous chapter that the type of expertise inculcated through the processes of making produces an essentially non-objectified form of knowledge which is neither propositional nor amenable to scrutiny. Secrecy, the second factor shrouding access to the builder's knowledge, has been interpreted as being an integral feature of the expert's customised discourse which is intentionally manipulated in order to tightly control the production and reproduction of customised persons within the trade.

The topic of intentionality is expanded in the second half of this chapter, and is divided into two major sections. In the first I propose a conceptual working model of intentionality based on philosopher Martin Heidegger's theory of time consciousness, and supported by contemporary neuro-philosophy, in order to frame subsequent discussions of both the patron's and the Master Builder's intention-driven choices and decisions regarding architectural enterprises. I argue that intentionality drives all decision-making processes at every level of our human existence, and that this process represents the mechanism which effectively structures our sense of temporality, as well as the continual present-ness of our experienced identity. The motivations of the project patrons for performing acts of *sadaqah*, or charitable endowments, are examined, and the effects of client-craftsman relations on the builder's intentionality are discussed. The degree of focussed intentionality attained by the master craftsman is likened to *ihsan*, the third and ultimate dimension of Islamic spirituality. It is illustrated that the harmonious balance achieved between the conceptual forces of imagination employed in the processes of design and making, and disciplined reason, is the qualifying characteristic of the *usta*'s expert status. The chapter concludes with a discussion of the current struggle between traditional builders and professionally trained architects in Sana'a over the control of the local building trade and the accompanying expert discourse.

It is important to clarify here my employment of such functionalist cognitive theories as physicalist-based explanations of body-mind knowledge, a modularity-of-mind hypothesis for understanding the division of labour between mental faculties, and a working model of human intentionality. These theories are not representative of the thoughts or beliefs that were held by the Sana'ani minaret builders regarding the manner in which they teach and learn, perform their trade-related tasks, or make their design and construction decisions. The theoretical premise of my study is firmly rooted in Western philosophical discourses about the mind which were entirely unfamiliar to my Yemeni co-workers; and, of significant note, my aim to better understand human behaviour, cognition, and agency in decision making does not consider the role or designs of God in these affairs. For many of my Muslim friends and colleagues in Yemen, even those who admitted to not being very devout in their religious observances, the relegation of God outside the scope of my analysis was curious, and in some cases was contentious. I have decisively chosen to address a series of questions about the human mind through an existing philosophical discourse to which I aim to make some small contribution, and build upon in my future studies. In restitution for any offence taken to this matter, I hope that my ethnographic representations of my work-mates, their lives and relationships during the time that I was there, and of their religious beliefs, social values and remarkably rich culture, are conveyed throughout this work as vibrantly as I experienced them and enjoyed them.

Chapter One

THE ʿADDIL MINARET

Reconsidering the Role of the
Mosque Minaret in Sanaʿa

When the minaret of the Masjid al-Shahidayn (located in the heart of the Sanaʿa souk) collapsed in the year 1302 of the Hejira (1885 AD), its structure was renewed by the distinguished Master of the Waqfs, Qadi Husayn bin ʿAli al-Amri. He told me that he spent one hundred and thirty riyals from the property of the Sanaʿa waqfs on its rebuilding, and the remainder was collected from the charitable merchants of the city. The collapse of the minaret had crushed the mosque's adjacent lavatory building where the Faqih ʿAbdullah al-Khabat had been performing his ritual ablutions, and resulted in his martyred death. During his lifetime, the Faqih, a learned man in the science of calculations, was regularly asked by the people (of the city) to foretell when the release from their suffering and hardship would come. He would respond: 'When the minaret of the al-Shahidayn is destroyed'.[1]

Since my childhood I've been enthralled with building. I have distinct memories of my parents lifting me up to the portholes cut in plywood barriers around construction sites so that I could peer into the vacuous foundations of new office buildings. From building blocks and crayon drawings I went on to study architecture at university and to design my own projects, but a persistent interest in the people who do the actual building led me to pursue studies in anthropology. Deeply inspired by the towering architecture of Yemen during previous travels in the Middle East, I chose to return to Sanaʿa, the capital city, in 1996 for field studies. That Spring I periodically passed by the site of a new minaret being erected in the former Turkish quarter of Bir al-ʿAzab to marvel at the efficient organisation of the building team and the precision of the masonry work. When I left Yemen for the Summer months, the foundations had been completed and the elegant black stone base of the tower was well underway.

By September, the square base was completed, rising fifteen meters above the street level, and a gold-painted, sheet-cut metal sign reading 'Ma Sha' Allah' (What God Wills) in calligraphic script hung from a thick chain above the door on the south side. The multi-faceted red brick tower which was being built above the base emulated the form and style of the minarets in the old walled city. The structure was reaching over thirty-five meters into the air before construction of its first calling platform had even begun. On one visit to the site, while standing in the street below, head tilted back and absorbed in watching the men precariously perched on the top of the walls laying bricks, I was approached by one of the older builders with a quizzical, almost suspicious, look on his face. He had observed me several times and was curious to know why I had been coming to watch them build. I responded simply that I admired their work.

'Who are you working for?', he asked.
'No one. I've come to Sana'a to study traditional builders', I replied.

His face softened a little with the faint trace of a smile, and he gestured to me to follow him.

We left the street to enter the forecourt of the mosque, and passed through the low doorway at the base of the minaret to our left. There were no slit windows in the thick masonry walls of the lower half of the base. Temporarily blinded in the darkness of the spiral stairwell, I groped my way along the cool face of the curved outer wall with my right hand, and climbed counter-clockwise up the uneven and debris-covered stairs, struggling to keep up with the old man's ascent. Slowly my eyes grew accustomed to the dark. The walls in the upper portion of the base and those of the brick tower above were intermittently pierced by slit windows, and I was soon able to discern the faces of labourers strung along the wind of the staircase. Nearly all greeted me as I passed, and several anxiously inquired "Where are you from?", "Do you speak Arabic?", "Are you a Muslim?". My brief and breathless responses were echoed by their shouts informing their fellow work-mates ahead about the nature of me, the foreign intruder.

The intensity of natural light grew stronger as we approached the top. In perfect form, the old man hopped from the last tread of the spiral staircase onto the top of the tower's circumference wall. He proceeded along the narrow top with several cat-like paces in the same counter-clockwise direction as our ascent, then paused and glanced back toward me with anticipation. I followed with carefully composed confidence, suddenly finding myself astride a brick wall, sixty centimetres thick with a drop into the stair-shaft on my left and a thirty five metre sheer drop on my right. It was incredibly exhilarating. A steady breeze cooled my perspiration-soaked T-shirt as I stood staring toward the horizons that surrounded us, spell-bound by the view over sprawling Sana'a. Once again the builder's face softened, and this time with a broad grin of satisfaction. It was clear to him that I was at ease at these heights. Apparently I had passed some sort of a test.

The following day I left Sana'a for the north of the country to scout out possible alternative locations for my study. Admittedly, however, it was the last impressions of my visit to the minaret which remained foremost in my thoughts. When I returned, I was determined to remain in Sana'a for the duration of my research. Despite my apprehension about the complications which might arise due to my being a non-Muslim labourer on a building site directly associated with a religious institution, I returned to the minaret in Bir al-'Azab in search of the old builder. As before, I positioned myself on the doorstep of the building across the road from the mosque and waited patiently for the opportunity to go back up to the top. It wasn't long before one of the labourers appeared, emerging from the forecourt carrying a shovel and a mesh screen for sifting the pile of gravel at the edge of the road. He recognised me immediately and we greeted one another. I expressed my interests to see the chief builders working at the top, and with a sweeping motion of his head in the direction of the entrance, I was on my second trip up the winding staircase. I carried a bucket of mortar passed on by another labourer on the stairs, and this time the message was passed on ahead that "the Canadian is coming". I felt confident and the darkness already seemed more familiar.

Surfacing back into daylight, I found there were now four builders working around the top, and I paused to look for a place to station myself out of the way of the activity. The old builder signalled me over to where he was laying bricks diagonal to, and slightly projecting over, the courses below, creating a jagged dog-tooth pattern. I knelt down next to him, gripped the edge of the precipice and leaned out-and-over to peer at the work on the underside of the projection. The builder explained how they were constructing the first balcony by progressively cantilevering the successive courses of brickwork. When he had finished, he paused before interrogating me once again about my presence in Sana'a.

'Are you an engineer?'
'An architect,' I replied, 'But I've come to Yemen to work with traditional builders, and ...', so as not to lose the opportunity, I quickly added in the next breath ' ... I'd be very interested in working for you. I want to learn about how you build in Sana'a.'

This time his grin was one of reassurance, and without any further hesitation he invited me to start the next morning.

I was now working for the Bayt al-Maswari and would remain in their service for the duration of my year-long field study. My decision to decline their offer of payment for my work was based on two accounts: firstly, I endeavoured to make as little economic impact on the project as possible and I did not wish to deprive any native builder of his potential, and much needed salary; and secondly, I was able to establish a system of reciprocity that was more favourable to my needs as a researcher. Refusing financial compensation enabled me to escape some of the confines imposed by contractual obligations and a rigidly defined hierarchy. I was therefore in a better position to make inquiries and to negotiate a somewhat less physically demanding work schedule than my fellow co-labourers. By maintaining my relatively ambiguous status within the structure of the work team, I was able to engage at various levels with a broader spectrum of builders, from labourers to the Masters, and thereby developed a more comprehensive understanding of the overall mechanics of the work team, the divisions of labour and the methods of passing on trade-related skills.

I Al-Khafanji's Poem

I quickly discovered that the al-Maswari family were esteemed Sana'ani craftsmen descended from the Master Builder, *Usta* 'Ali Said al-Maswari, and the current minaret project was being directed by two of his surviving sons, Muhammad and Ahmad. The success of the family's first minaret commissioned in 1980 led to their subsequent building of more than twenty of these towering structures by the time of my study in 1996. Unquestionably, the Bayt al-Maswari and their patrons had been largely responsible for the renaissance of lofty minaret towers that have pierced the skyline of the newer city quarters in the last two decades. I also soon realised that the mosque for which the current minaret was being built, the Masjid 'Addil , was the same 'Addil Mosque personified as the central character in 'Ali ibn al-Hasan al-Khafanji's celebrated poem on the Mosques of Sana'a.[2] The mid-eighteenth century poem, written in a style known as *humayni* verse and using a modified form of Sana'a's distinctive dialect, was translated and published in German with commentary by Harald Vocke in 1973,[3] and in English by R.B. Serjeant in 1983.[4] However, it was through my own initial familiarity with the latter text and subsequent verification with several learned Sana'anis that I made the connection. Not surprisingly, the builders, with limited education and low levels of literacy, were unfamiliar with this famous and somewhat prophetic poem.

Al-Khafanji's poem of the Mosques of Sana'a depicts an inspection visit paid by the Great Mosque of the city to the 'Addil Mosque which is situated in the former Turkish suburb of Bir al-'Azab and near to the Bab al-Balagah, the gate to the old Jewish quarter. The Great Mosque is portrayed by the author as a character of wealth and distinguished status, and the impoverished Masjid 'Addil makes a desperate plea to him for improved conditions of his neglected state. From the poet's mid-eighteenth account it would seem reasonable to assume that the condition of the 'Addil Mosque at that time closely resembled the description of the 'old mosques' rendered by al-Khafanji's contemporary, the author Sayyid Jamal al-Din. In his *City of Divine & Earthly Joys: The Description of Sana'a*, Jamal al-Din writes:

The mosques [of old] had neither courtyards, ablution places with running water, lavatories, nor gardens – indeed they were devoid of such attractions, and the most one could hope for was the odd one might have a well (and this without any well-gear – not even a hand-raised bucket).[5]

By personifying the Mosques of Sana'a in a poetic form that combines both delightful charm and satire, al-Khafanji was able to safely compose a political commentary on the social hierarchy, on the daily life and customs of his fellow citizens, and on the position of the mosques in his city without risking his life and freedom. Al-Khafanji made regular use of his poetic licence to remark on the local political and social struggles of his time. In one poem, he composed a dispute between two principle *makharif*, or suburban resorts of Sana'a, al-Rawdah and Bir al-'Azab, which he personified as bickering women, and Sana'a itself intervened in order to resolve the quarrel.[6] Harald Vocke, in his interpretation of the poem on the Mosques of Sana'a, suggest that the author's politics expressed in the work were directed at the inspector of the mosque *waqfs* at the time, Qadi 'Abdullah al-'Arasi who, like many of the *nazirs*, may have been corrupt.[7] Serjeant conjectures that the poetic motives were more generally aimed at complaining "of the poverty and neglect of some Sana'a mosques". I have transcribed al-Khafanji's poem in its entirety below as translated by Serjeant, and following, I will address those verses which directly concern a discussion of mosque minarets.

1 Says 'Addil , '*Wa-'l-salam*, a greeting
   Proffered to the Jami' in sincerity'.
2 When he arrives I'll lay the case before him.
   'Addil says he's got a complaint to make,
3 'First let him build for me ablution-places,
   Second, have them clean out the well.
4 The wind (chills) the foot-pool, blowing like bellows.
   I have no lamp to light me in the evening,
5 My east wall leans over, like to fall,
   God preserve it from worse ill-fortune.
6 No friend have I, none to make me known,
   On this question Qarish, ask al-Ashrafiyyah!
7 If you've anything, your friend's the one on whom to spend it.
   Oh Jami', if only – when you honour me by visiting
8 To get to know and verify the state I'm in –
   You'd favour a fellow, with your kindly face.'

9   The Jami', turning to him, said 'Why not?
        Already people have acquainted me with your condition.
10   Assuredly! You're welcome – full willingly.
        To come to Bir al-'Azab is little trouble to me.'
11   Till the Friday of Rajab he tarried, then, of a sudden, came.
        Out from the Wall he went, passed through the open countryside.
12   On his right Madrasah and Abzar.
        Resplendent as the moon was his appearance.
13   Hearing each from the other of his venture forth, the Mosques
        Came running up to join him with the still prostrate in prayer.
14   On went the Jami', bearing all worshippers with him,
        Replete with learning, his bearing stately,
15   Came to al-Juruf, sat himself down,
        Had Madhhab and Abhar serve him with refreshments there.
16   Up with his plaints arrived Nizayli, 'Aid me', he cried, '*Ya gharatah*!'
        A present he brought – a gift of seductive musk,
17   Came into his presence, flung down his mats.
        Said the Jami', by distress affected, by sighing overcome.
18   'On you Bounteous God will cast his regard, (pray patience),
        Our brother Yaqut has (somehow) mislaid the waqf deeds.'
19   Said 'Addil , interposing, 'My brother, wait,
        Allow the Enlightened Jami' to rest and listen,
20   Then speak, and there be some point to your word,
        Else brother, all becomes a mob and confusion.'
21   Said Nizayli, 'I am endowed with gifts
        Of incense, cupboards for books, sweepers' wages.
22   Both plain and (mountain) shortcuts have I travelled.
        My ablution places are not keenly cold in winter.'
23   The Jami' turned, taking the road to 'Addil .
        Hanzal, in new clothes had come up meanwhile,
24   Qadi approached (to join them) at a trot,
        And after them, Bahmah, rather more slowly.
25   The apricot trees joyously trilled their welcome to the Jami',
26/1 To him the pear-trees waved their branches,
        The cranes flapped and clapped their wings.
26a   'Addil , poor fellow, was overcome with joy,
        From the moment his court appeared, kept exclaiming, 'Power lies
            with God alone',
26b   Eyes tense with expectation, kept coming back to the Jami'.
26/2 'One ounce (of lamp oil) a week have I', he said,
27   'Never, in my time, have I known intendant's (care),
        Nor ever (spared) they two hot coals to warm me,
28   But if no waqf I have, you are yet Protector to me,
        Since, oh mighty Lord, you remain my sole resource!'

29 The Jami' replying to him said, 'Be of good cheer.
        We must attend to you, see to your case,
30 Mend those parts of you in disrepair, re-build them.
        You and I will split the catch-crop when it comes in,
31 Carpets I'll give you from my northern wing.
        This carpet, no-one can deny, is done.
32 You, my lad, are foremost in my mind,
        Yes, by the Prophet, you're in a sorry plight.'
33 Up came Sayyad, in his haste without belt and dagger,
        Running all distraught from the door of Hanzal,
34 A long sheet of paper, gummed, in his hand,
        A fair copy wherein were matters set forth clearly.
35 Greeting the Jami', he went on to say, 'Listen,
        This poor fellow needs you – he is destitute.
36 If you can spare something and be generous
        Will you from your surplus, spare him just a little?'
37 The Jami' made reply most satisfactory,
        'In the past', said he, 'I've been neglectful of you.
38 I shall send to Qas'ah, and I'll get al-Hifafi
        To estimate what you'll have to pay in men's wages.'
39 Up rose 'Addil , responding with a blessing (on him),
        'May the Lord guard you from the glances of the evil.
40 With these words of yours you have relieved my heart,
        For you are of those versed in the law, sensitive of their honour.'
41 The Jami' turned to him, said laughingly,
        'A splendid thing to have relieved your heart!
42 Negotiations to pave your court with someone I've begun already,
        And to extend it westward of the building,
43 Since you are needy and must have more revenue.
        Your prayer-niche is tinier than a leather bag!
44 They say that worship in you is acceptable (to God),
        Yet no (waqf) do you own of land, or hostelry.'
45 Said Mu'id, 'I saw good fortune for you in a dream.
        That you were standing beside the door of Abu Tayr
46 When Abu 'l-Khayr stopped in your prayer niche
        And the minaret that belongs to you was like Khuwarnaq.
47 On waking, I betook myself to the dream-interpreter,
        And there was al-Yazidi sitting in Zumur -
48 A faqih who knows all things have One who disposes -
        Compared to him Ibn Sirin doesn't come up to here!
49 To him I said, "I've seen a vision
        That 'Addil has come into conjunction with Pleiades,
50 That he is full set for fortune
        And has courts like those built by 'Amir".'

51   Said (the Jami'), 'This is the best of the things you've seen.
      If what you say be realised you have (indeed) (re-)built him.

52   Summoned him to well-being and prosperity.
      Favourable regard must be accorded him.

53   His position is beside the Enlightened (Jami').
      He must receive favourable consideration, be re-furbished,

54   His prayer-niche decorated with green *laz,*
      He must be transformed into a veritable Jami' of the Banu Umayyah!

55   He must have the field of 'Iyal Jassar,
      In which they'll plant onions and (fruit) trees.

56   There they'll study al-Bayan and al-Azhar,
      Do the Mulhah and the Shatibiyyah.

56a   A book-cupboard they'll make at his prayer-niche,
      A store too, they'll make for the waqf.

56b   Oil (for his lamps) they have assured to him,
      He shall have an intendent, strong of authority,

57   In the minaret they'll plant a muezzin
      Whose prayer-call will leave Yashu''s ear in the synagogue astoundingly ringing,

58   So resounding his 'Subhan Allah' it'd make the Jinn fart.
      An Imam he'll have, to lead the Prayer, a man of religion, with a stipend.'

59   The Jami' took his stately way to Sana'a,
      'Addil accompanying him, running by his side.

60   God preserve and protect his fair and noble brow,
      For gracious was ever of his nature.

61   His way took him opposite Shamlah,
      Passing Shu'ub north (of the city) on his left.

61a   The apricot trees joyously trilled their welcome to the Jami',

61b   The cranes flapped and clapped their wings
      The pear trees shook their leafy (stems),
      The lips of the glistening orchard, in a smile, half parted.

62   A deal of kindness had he done, and charity,
      With kind intent attended to his purpose.

63   Onward he proceeded till he'd passed by the Khanadiq,
      (Where) he recalled a word he'd had earlier with al-Bustan

64   That he is poverty stricken, the most fortunate evening being
      When he has a companion within the walls; he's like an open yard.

65   At once Farwah addresses herself to him,
      Waving flags, flag-poles from Shu'ub.

66   'My palm-leaf mats have had a long life,' said she,
      'As anyone who reclines (upon them) knows.'

67   Mashhad beside her kept a-calling,
      'How long have I been waiting for the prayer?' she said.

68   To her the Jami' turned his attention, saying, 'Wait,
        The best things are those that are slow in coming.'
69   As he entered, the city trilled for him cries of joy,
        Proceeding calmly and with courteousness.
70   On the ground of Azal (Sana'a) may festival ever continue,
        And for all time its rank remain exalted.[8]

II Reconsidering the Role of the Minaret in Zaydi Yemen

Of the poem's seventy verses, several raise important questions regarding the status of the minaret as a building type in the historically Zaydi-dominated city of Sana'a. Zaydism, a moderate sect of Shi'i Islam that traces its Imamate through Zayd ibn 'Ali ibn Husayn ibn 'Ali ibn Abi Talib who was martyred during the reign of the Umayyad Caliph al-Hisham, has courted contentious views of the tower minaret since the tenth century AD. Like the spokesmen of several other orthodox Islamic communities, some Zaydi theologians have advocated avoidance of building minarets on the purist grounds that such structures are ostentatious and distract from spiritual focus, and Lewcock reports that opposition to minarets was based on the grounds that "they made mosques resemble churches".[9] Nevertheless, many have been, and continue to be, built in Sana'a. With reference to several of the poem's verses I will reconsider the role of the minaret as component of Zaydi religious architecture in Yemen.

45   Said Mu'id, 'I saw good fortune for you in a dream.
        That you were standing beside the door of Abu Tayr
46   When Abu 'l-Khayr stopped in your prayer niche
        And the minaret that belongs to you was like Khuwarnaq ....
51   Said (the Jami'), 'This is the best of the things you've seen.
        If what you say be realised you have (indeed) (re-)built him.
52   Summoned him to well-being and prosperity.
        Favourable regard must be accorded him ....
57   In the minaret they'll plant a muezzin
        Whose prayer-call will leave Yashu''s ear in the synagogue
        astoundingly ringing,
58   So resounding his 'Subhan Allah' it'd make the Jinn fart.
        An Imam he'll have, to lead the Prayer, a man of religion, with a
        stipend.'

'Yashu'' in verse 57 refers to the Jews of the Qa' al-Yahud (The Jewish Quarter of Sana'a) located immediately to the west of the 'Addil Mosque. If al-Khafanji is suggesting in that same verse, speaking through the voice of the Great Mosque, that a *muezzin* (prayer caller) will be planted in the future minaret envisioned by the unidentified character 'Mu'id' in his dreams (verse 46), then it is rather ironic that there was not a single Jew left residing in the Qa' al-Yahud by 1996 to hear the "resounding Subhaan Allah" crackling over the loudspeakers of the new, and long-awaited minaret. However, the Arabic of the poem, as transcribed by al-Hajari,[10] employs the word *sauma'a* which Serjeant perhaps questionably translates here as minaret. Though in Yemeni spoken Arabic the term *sauma'a* has come to be used interchangeably with *menara* (minaret), *sauma'a* in literal translation is rooted in the pre-Islamic definition of 'a monk's cell or hermitage'. During the early Islamic period, the word became associated with the small cabin-like structures built at the top of the staircase to a mosque roof which served to protect the *muezzin* from intense sun and poor weather conditions. Therefore our reading of the poet's use of *sauma'a* leads to an ambiguous interpretation, particularly in verse 57.

Despite the Arabic use of *sauma'a* in verse 46 as well, the metaphorical reference to 'khuwarnaq' used to physically describe Mu'id's envisioned minaret for the Masjid 'Addil more clearly indicates that al-Khafanji had in mind a towering structure, and not a cabin at the top of the stairs. Both Vocke and Serjeant agree in their analyses that the name 'Khuwarnaq' pertains to the legendary, richly decorated, and presumably towering palace of King Bahram, and I will address this reference more fully later. In verse 57, however, it is not clear whether the Jami' Mosque is referring to Mu'id's envisioned towering minaret or whether a *muezzin* will be planted in an already existing cabin-like structure. It is certain that the 'Addil Mosque did not have a tower-style minaret before 1996, and prior to this date there had been a metal tripod structure erected on the roof of the mosque which supported a loudspeaker for broadcasting the call to prayer. Before the advent of the loudspeaker, and at the time of al-Khafanji's writing, it is very possible that there existed a *sauma'a*, in the standard Arabic sense, on the mosque roof, and it is in this edifice that the Jami' Mosque of the poem intended to "plant a *muezzin*".

The call to prayer (*adhan*) in Islam, popularly thought to have been invented in the Hijra or the following year (623–4 AD) in response to the Jewish use of the horn and the Christian use of wooden clappers,[11] or bells,[12] for summoning worshippers, was indeed originally made from either the courtyard of the mosque or the roof. This practice continued well into the twentieth century in Yemen, and even at the time of their field studies in Sana'a, Serjeant noted that:

> the call to prayer is mostly delivered in the court (*sawh*) of the mosque, even in the mosques which have a minaret.[13]

They name only three mosques aside from the Jami' Mosque where the call to prayer was being delivered from the minaret, thus strongly supporting the view that the minaret is not an essential component of Islamic religious ritual. As of this writing, many of the mosques in Sana'a are still without tower minarets including several notable mosques such as the al-Mutawakkil Mosque built by Imam al-Mutawakkil in 1726 and extended during the reign of Imam Yahya in 1936. However, in recent years many existing mosques, like the Masjid 'Addil, have been endowed with minarets, and plans for a seventy meter high tower for the al-Mutawakkil were in the pipeline at the time of my study.

The Poem as Prophesy

Al-Khafanji's foretelling of a minaret through the voice of Mu'id in verse 46, and the notion of "re-building him" reinforced by the Jami' in verses 51 and 52, are of interest in several respects. Firstly, from historical records we know that there was a considerable amount of new construction and renewal commissioned for religious institutions in Sana'a in the period from the mid to late eighteenth century. Of note are Imam al-Mansur's benefaction of a minaret to the Masjid Musa in 1747–48 AD, and his extension to the Mosque of al-Abhar; the renewal of the Mosque of Nusayr in 1748; the building of the mosque and minaret of Qubat al-Mahdi 'Abbas by the Imam of the same name in 1750–51; and the erection of the Zumur minaret in 1790.[14] The 'Addil Mosque was indeed renewed by Imam al-Mansur around the turn of the eighteenth century (1224 AH), and later, in the mid-nineteenth century (1266 AH), a *saqifah* (shelter) was built over the courtyard of the mosque by the reigning al-Mutawakkil Imam. In the mid 1990s (AD), the finances for the complete rebuilding of the 'Addil Mosque and the addition of a stone and kiln-baked brick minaret of over fifty meters in height was endowed through the charitable funds made available by the Bayt Fahim, an established Sana'ani industrial and trading family. Amusingly, Mu'id's "dream" prefigures the advent of that minaret tower by nearly two and a half centuries.

The Power of Height: verticality in secular and religious architecture

Secondly, returning to the poet's metaphorical reference to 'Khuwarnaq', the palace of King Bahram, the comparison of the religious edifice with a legendary example of secular palatial architecture provokes some consideration of a point made by Jonathan Bloom, a leading theorist on the origin of the minaret. In his historical and contextual reconstruction of the minaret's development as a "universal architectural symbol of Islam",[15] Bloom asserts that prior to the ninth century in the Islamic Middle East, height had been associated with secular, palatial architecture. However, an increasingly regular attachment of towers to mosques under the 'Abbasids, and not under the Umayyads as had been previously advocated by Creswell,[16] led to a shift in the semiotics of architecture: height became associated exclusively with religious building in Islam.[17] Al-Khafanji's comparison, presumably premised on the vertical aspects of both structures, as well as perhaps the quality of decorative aesthetic, would seem to support Bloom's hypothesis that the attribute of height was transferred from secular to religious edifices. However in Yemen, like very few other regions of the Islamic Middle East such as the Ksar in Morocco, height is an attribute which has been historically shared by both building types, and arguably constitutes a sign of power and affluence in both cases.

Throughout much of South Arabia, there is a close analogy between the height of the dwellings, or tower houses, and the associated prestige of the owners, many of whom continue to the present day to commission such buildings. The existence of tower houses in Yemen, often rising five and more stories, dates back to at least the advent of Islam and very likely several centuries earlier as evidenced by the stelae of Axum,[18] and archaeological evidence from Shabwa, the capital of Ancient Hadramawt.[19] Ibn Rustah provides us with the earliest description of Sana'a houses in Islamic times (903–913 AD), writing that the city is populous and with 'fine dwellings, some above others ...' (in reference to their height).[20] Jamal al-Din in his engaging eighteenth-century depiction of the tower house which describes their height, quality of craftsmanship, and windows of coloured glass, remarks that:

> Such, then, are the houses of the rulers and their assistants and followers. The form is imitated in the houses of the wealthy and the merchants, then [in turn] of those who come after them [in the social order], in the houses of the artisans and farmers who follow them, and in the houses of the paupers.[21]

It has been postulated that for millennia the commissioning of splendid buildings has been one of the few available, yet most effective means for the individual in Yemen to express his or her personal success, and thus explains the 'distinction attained' by the region's architecture.[22] In contrast with the characteristic urban dwelling types of Islamic North Africa whose reserved exterior appearance and inward-looking spatial arrangement have developed in conformity with ideals of privacy set forth by the Maliki School of interpretation, the exteriors of Sana'ani residences are extroverted displays of wealth and social status, and the positioning of rooms is planned in accordance with their assigned importance. The most significance social spaces, such as the *mafraj*, are located on the uppermost stories and address the most desirable direction in terms of light, air circulation, and view. It is a well-known formula in Sana'a that a house whose principal facade faces south is a complete house; one that faces west is half a house; one facing east is a quarter house; and one facing north is no house at all[23] – though practical conditions do not permit strict adherence to this prescription.

Like the tower house form, minarets in Sana'a and elsewhere in the Islamic world have served as markers for the wealth and status of their benefactors in addition to symbolising the Islamic community. One of Bloom's most important contributions to our understanding of the minaret is his suggestion that the attachment of a tower to a mosque and the provision of a distinctive place for the call to prayer were initially unrelated ideas that became conflated at a much later date.[24] Although the tower may have been used for the call to prayer as early as the ninth century, it only became more firmly associated with that function in the eleventh century. He argues that the 'Abbasids introduced the tower into the mosque 'as a symbol of an ideology', and once out of 'Abbasid control and in the hands of other empire builders in the expanding lands of Islam, the minaret was given form, proportion and aesthetic expression that 'reflected doctrinal variation'. Architecturally, the tower had always effectively 'signified the power of the individuals or the institutions that inhabited it or controlled it'. Following from this, Bloom asserts that the meaning of the minaret is therefore not derived from 'its purported function – the call to prayer – nor from its association with any religiously significant form'.[25] He concludes that the belief that 'the loudspeaker spelled the functional obsolescence of the minaret' is unfounded: 'the loudspeaker has nothing to do with the tower'.[26]

Several factors that have been observed in Sana'a support this thesis. As noted earlier, prior to the installation of loudspeakers on the domes or in the upper-level openings of the tower minarets in Sana'a, the call was 'mostly delivered in the court ..., even in the mosques which [had] a minaret'.[27] Secondly, it should be recognised that the majority of calling platforms on the minarets of Sana'a are situated at a level too high above the streets of the dense urban environment for the *adhan* to be effectively broadcast to a listening audience below without mechanical amplification. A similar observation about the functional limitations of minarets with regards to the call to prayer is made by Robert Hillenbrand in his study of Islamic architecture.[28] Finally, an additional point for consideration, which I address more fully in chapter four, is the roles of both the patron and the Master Builders in the determination of the form and height of the minaret towers.

Traditionally, ruling Imams, and increasingly wealthy merchants, made munificent bequests to religious institutions through acts of official patronage or *sadaqah* (charitable alms). In Zaydi Yemen, the *madrasah*, or religious school or college characteristically affiliated with the mosque, and traditionally a prominent object of patronage throughout much of the Muslim world, was essentially non-existent.[29] Therefore the commissioning of new mosques and renewal of existing ones, the building of ablution places, the paving of courtyards in stone, and the erection of minarets became the primary mediums for the conspicuous display of power, as well as religious devotion. With few exceptions, the financial source of the Bayt al-Maswari commissions since 1980 have been charitable trusts set up by prominent individuals in the city, and in some cases supplementary funding has come from the government Ministry of *Awqaf*. For instance, the projected minaret for the al-Mutawakkil Mosque was to be financed in equal parts by the trust of a Sanaʿani merchant family in the name of their deceased head and by the government. Construction materials, styles, and the height of the towers were negotiated before the commencement of the projects, and were largely determined by the available wealth of the client, as well as by the artistic objectives of the builders to attain what they consider to be aesthetically pleasing proportions. On at least one occasion, government-imposed regulations restricted the height of one of the al-Maswari's structures due to its proximity to a security sensitive zone. Normally, the aim of both patron and builder was to build as high as possible with the available resources, not unlike the commission for a traditional Sanaʿani urban tower house.

The above points invite a reconsideration of both the standard etymology and the significance of the minaret in Sanaʿa. Clearly these structures were not erected primarily to service the function of the *adhan*, but rather they are markers in the urban landscape imbued with symbolic properties. The Arabic terms most regularly used to denote the minaret are *miʾdhana*, *saumaʿa*, and *menara*, and each describes a distinct feature of this building type. The first, *miʾdhana*, is derived from *adhan* (call to prayer) and describes the place where the call is made. Although it directly relates this ritual function to the building, it is not commonly used, and this may reflect the fact that the minaret was not habitually used for making the call to prayer in Sanaʿa. Hillenbrand equally suggests that the comparative rarity of the use of *miʾdhana* in the Arabic-speaking world may indicate that the "early minaret, later tradition notwithstanding, did not serve primarily (or perhaps at all) as a place for [this function]".[30]

Some theorists have drawn a connection between the literal translation of the word *'menara'*, a 'place of light', with the function of the call to prayer through the metaphorical correlation between the concept of 'illuminating light' and the 'Word of God' which is spread in the call issued from the tower.[31] Again, however, this leads mistakenly to the functionalist conflation of the *adhan* with the original meaning of the minaret.

In Sanaʿa, like North Africa, the tower minaret is still most popularly referred to as "*saumaʿa*" for which the etymological root was explained above, and in more standard use denotes the box-like projections built on the mosque roof. It has been suggested that the tower ultimately derived from the gradual heightening of these structures,[32] but there is no evidence to suggest that the characteristic Sana'ani minaret, usually built as an independent structure from the mosque building, was the outcome of any such evolutionary process. In fact the box-like form continued to be produced contemporaneously with tower minarets into the twentieth century and, like the finely crafted one on top of the al-Mutawakkil Mosque, is regarded as a distinct structure from the tower. Perhaps the peculiarity that both types of structure share the same terminology can be simply explained by the fact that both contain a 'cell' for harbouring the *muezzin*. Whether or not a minaret in Sana'a were built for making the call to prayer, all accommodate the characteristic calling platform reached by an interior spiral staircase. Nevertheless, it would seem that in reference to the above discussion on the role of patrons and builders, *'menara'*, interpreted as being "a word ultimately derived from an Arabic word meaning 'sign' or 'marker'", is ostensibly the most appropriate term for describing these structures.[33] Bloom concludes that *menara* is the most appropriate term for the universal architectural symbol of Islam, while Hillenbrand notes that "the term manara was applied, by a familiar process of semantic depreciation, to sign-posts, boundary stones and watch-towers even when no particular association with light or fire was intended".[34]

A similar precedent of the tower-as-'marker' in Yemen, though not Zaydi, is the tower-like projections built by the Rasulids (1229–1454 AD) above the *mihrab*s in their mosques in Taizz. These mosques were originally built without minarets and the finely decorated *mihrab* tower served primarily to sign the location of the prayer niche along the *qiblah* wall,[35] as well as the power of those that built them. I contend that the minaret towers of Sana'a are symbolically representative of three salient factors: firstly, the location of the associated mosque institution and the centre of its congregation, or more generally, Islam; secondly, the status of the patron(s) whose intentionalities have been driven by a combination of both religious and secular motivations; and finally the skill and dedication of the Master craftsmen who built them. Though this latter point is not addressed in other studies on minarets, I maintain that it is an important aspect of the symbolic display, and it constitutes a central focus in my study. The multi-vocal approach to the symbolic analysis of minarets which I have taken responds to Hillenbrand's caution against "a too narrowly political or secular interpretation of the minaret".[36] Referring to numerous and lofty Seljuq minarets of Isfahan, he rightly notes that these edifices are expressions of both "conspicuous consumption and conspicuous piety".[37] In a similar vein, Marcicq and Wiet in their study of the Djam Minaret built by the Ghorid prince Ghiyath al-Din (1153–1203 AD) point out that:

En exaltant la grandeur du souverain qui l'a érige, ce monument reste d'ailleurs fidèle à son caractère religieux: ne proclame-t-il pas que: 'Le secours (qui mène à la victoire) vient de Dieu'?[38]

41

Zaydis and Minarets

Returning to al-Khafanji's poem on the Mosques of Sana'a, an important point gleaned from the six verses addressing minarets is that the poet made such a pronounced and agreeable reference to these tower forms at all. This inspires a more careful consideration of the Zaydi position vis à vis mosque minarets in Yemen. Al-Khafanji's distinguished Zaydi lineage can be traced to Imam al-Qasim bin Muhammad, also known as Imam al-Qasim al-Kabir (The Great), a national hero who led rebellious offensives against the occupying Ottomans in the early seventeenth century and is regarded as the founder of modern Zaydi Yemen. During the ensuing period of Zaydi government that lasted into the mid-nineteenth century it is reported that few mosques were built with minarets,[39] generally interpreted to be a reflection of Zaydi opposition to these towering structures. This not always apparent opposition is historically rooted in a tenth-century decree of the Zaydi Imam al-Utrush. Al-Utrush was an enlightened and militaristic Imam of the Zaydi state of Tabaristan (now Northern Iran) on the Caspian Sea, and ruled contemporaneously with al-Hadi ila 'l-Haqq, the first Zaydi Imam of Yemen. In his treatise on *hisba*, or market regulations, he pronounced that "mosques must not be made like churches, contain pictures, be ornamented with gold, hung with curtains or decorated with plaster work",[40] and:

> The minaret of mosques should not be raised above its roof, and those that are higher, should be made lower, for they report that the Commander of the Believers, 'Ali, God's blessings upon him, said, 'Do not raise the menara of the mosque above its wall. And its structure – what is attached to it – is to be equal in height to the roof of the mosque.'[41]

However, Bloom points out that the use of the term *'menara'* is clearly anachronistic, for the word did not exist in the context of mosque architecture when 'Ali, the Commander of the Believers, died in 41 AH/661 AD, and thus calls the reliability of the passage into question. He proposes that, more likely, these Zaydi codes had developed in political response to the introduction of the mosque tower under the adversarial government of the 'Abbasids. More directly, the Zaydi theocracy of Tabaristan was threatened first by the appointed governors from Baghdad who ruled over the surrounding provinces of Central Asia, and from the ninth century onwards by the succeeding dynasties of the Tahirids (821–73 AD) and the Saffarids (867–963), all of whom had accepted and proliferated the tower minaret form.[42] Bloom remarks in a footnote to his argument that 'legally, the Zaydis termed them [minarets] 'reprehensible' (*makruh*) and had them destroyed', and questionably resolves that a *menara* was 'inimical to a Zaydi mosque'.[43]

Perhaps this general assumption could be applied to the political motivations of the short-lived Zaydi community on the Caspian Sea which felt more direct competition from their Tahirid and Saffarid rivals, but the reasoning would not have likely endured in the distant region of Yemen so far removed from the centres of 'Abbasid power and the Central Asian Dynasties of the ninth and tenth centuries. While it is true that Sana'a served as the seat of the 'Abbasid governors in South Arabia, and occasionally their control extended south to al-Janad and the Hadramawt, their influence was already weakened following the rise of the local Yu'firid Dynasty in 847 AD, and soon dwindled to the point where they could no longer appoint governors in the region. The first Zaydi Imam, al-Hadi ila 'l-Haqq, only settled in Yemen some fifty years later. Born in Medina in 859–60, he was "fired by the ambition to rid the Yemen of all evil practices and to bring to her people the benefits of his own version of Islam".[44] One of his first challenges at the turn of the tenth century was indeed to subdue the unruly Al Tarif – clients of the troubled Yu'firids – and the Khafatim – the Turkish soldiery brought by the then departed 'Abbasid governor 'Ali b. al-Husayn.[45] With the exception of this event there was seemingly minimal overlap between the political spheres of the Iraqi 'Abbasids and the Yemeni Zaydis, and it therefore seems highly improbable that any apparent adherence to al-Utrush's doctrine regarding the height of minarets could have been motivated by the community's political interpretation and contestation of these tower forms.

Al-Hadi returned north to the town of Sadah in 902, leaving Sana'a to its own fate, and consolidated Zaydi power in the north of the country where it remained until Imam Salah al-Din made Sana'a his seat of control in 1381 AD. From that point onward, Sana'a was the focus of Zaydi state interest, while Sadah has remained the sect's spiritual centre. If during their intermittent periods of rule there was indeed a politicised resistance to minaret building amongst the Zaydi community, and couched in codes of religious observance as Bloom has suggested, then more likely it would have been directed against such rival regional powers as the Ziyadids, Sulayhids, Rasulids, or the occupying Ottomans. These groups were either Sunnis or Isma'ilis, and all commissioned minarets which were stylistically character-istic of their eras and rulers. As Barbara Finster reports in *An Outline of the History of Islamic Religious Architecture in Yemen*, the minarets produced there in a variety of forms are "original creations peculiar to the country",[46] and the author cites examples of the Ziyadid minarets of Zabid, the "incomparable towers" of the Rasulid Ashrafiyyah, as well as the brick minaret of the palace mosque at Zafar Dhi Bin built by Zaydi Imam al-Mansur Billah (ca. 1200) (see plates 4–9).[47]

In fact, although it has been noted that most mosques in Yemen were built without minarets during the period of Zaydi government, several significant ones nevertheless were, and there is no historical indication that

PLATE 4
Minaret in Zabid.

PLATE 5
Minaret of Zafar Dhi Bin. Detail of
decorative brickwork resembling
snakes.

44

PLATE 6
Minaret in Jibla.

PLATE 7
Minaret in Kawkaban.

45

PLATE 8
Minaret of the Great Mosque in Sadah.

PLATE 9
Minaret in Mukallah.

PLATE 10
Minaret of the al-Madrasah Mosque,
Sana'a (1519–1520 AD).

PLATE 11
Detail of the brick base of the al-
Madrasah minaret.

existing minarets were "made lower" by the Zaydis in Sana'a in accordance with al-Utrush's decree. Of prominent note is the minaret of the al-Madrasah Mosque in an eastern quarter of the old walled city (see plates 10 and 11).[48] Al-Hajari records that it was erected in 926 AH/ 1519–20 AD by Imam al-Mutawakkil Yahya Sharaf al-Din,[49] and this was during a period of reinstated Zaydi rule following the demise of the occupying Mamluks. This minaret is of particular interest as it is thought to be the earliest example of its kind in Sana'a, and its form, proportion and decorative brick style set the precedent for traditional Sana'ani minarets to the present day. The al-Madrasah minaret, entirely constructed in kiln-baked brick and embellished with decorative and architectural motifs in relief from bottom to top, is comprised of a rectangular plinth which supports a columnar shaft capped by a projecting balcony, and an octagonal tower of smaller diameter above which is crowned by a fluted dome. It will become evident in the following chapters which concentrate on the contemporary building practices of the Bayt al-Maswari that their minaret towers conform to this formula with the exception that the plinths are constructed in stone rather than brick.

Other notable tower minarets built in Sana'a during periods of Zaydi government before the beginning of the Revolution in 1962 include that of the Masjid al-Abhar commissioned by the wife of Imam Salah al-Din in 1374–5 AD; Imam al-Mansur al-Husayn b. al-Qasim's commission for the addition of a minaret to the existing Mosque of Musa in 1747–8 AD (see plate 12), which the Bayt al-Maswari profess to be the precedent for their own constructions; and the typical Sana'ani proportioned minaret belonging to the Ottoman-style Qubbat al-Mahdi 'Abbas (see plate 13) built by the Imam of that name in 1750–1 AD. During his brief official visit to the Imam, Robert Finlay, Residency Doctor at the British factory at Mocha, counted eight mosques with tall minarets, although there were clearly more.[50] An ongoing practice of architectural patronage for religious institutions in the twentieth century is evidenced by Carl Rathjens and Hermann von Wissman in their 1920s study of Sana'a in which they remark that "every architectural resource is used to beautify the mosques, and above all their minarets, called *sauma'a*".[51] More recently, the minaret of the Zumur Mosque (see plate 14) located close to the Bab [al]-Shu'ub (the northern gate of the walled city), originally erected by Imam al-Mansur in 1790–1 AD, was heightened and adorned with its unique fretwork brick patterning in 1945 during the reign of Imam Yahya; and the minaret of Masjid Ibn al-Husayn was also built in 1936–7 during Yahya's era[52]

Local Zaydi authors paid tribute to these soaring structures that pierce the city's skyline of towering houses in their urban and architectural accounts. Al-Khafanji's contemporary, Sayyid Jamal al-Din, wrote in his eighteenth-century description of Sana'a that:

PLATE 12
Minaret of the Mosque of Musa,
Sana'a (1747–1748 AD).

PLATE 13
Minaret of the Qubbat al-
Mahdi 'Abbas, Sana'a
(1750–1751 AD).

49

PLATE 14
Minaret of the Zumur Mosque,
Sanaʿa (1790–1791).

PLATE 15
Concave faience tiles set in
carved decorative plaster.

It is a fact that many of these mosques have minarets rising high into the air. Some of these minarets have decorations and inscriptions [executed in] burnt brick and gypsum-plaster, and various types of glass discs set into them. The like of these minarets is not to be found in the Two Noble Holy Cities, according to what we have been informed by more than one informant.[53]

It is worth noting that the 'glass discs' in the passage refer to faience tiles which are found on the base of the minaret of al-Abhar, as well as on the minaret at Zafar Dhibin. Finster notes that "these may be understood as distant echoes of Anatolian-Iranian architectural ornamentation, which is used only in Zaydi cult buildings in Yemen" (see plate 15).[54]

Other authors sang their praise in verse, such as the poem composed by Qasim b. Yahya al-Amir for the minaret of the Masjid Musa which praises both the beauty of the edifice and the piety of the patron, Imam al-Mansur al-Husayn b. al-Qasim:

Oh! How lovely is the minaret
    Towering over all the buildings
Indeed it has benefited the person who built it
    Glory, Reward, and Praise
I mean by this that al-Mansur
    Our Master, the beautiful al-Husayn
Who preserved the whiteness and
    the brownness of the heights of Yemen
Satisfaction of a chronicler
    Indeed who has gained a good reputation.[55]

After the revolution and the elimination of the Imamate, and the subsequent rapid expansion of the city and diversification of its population, there has been a profusion of mosque and minaret building in and around the capital. As I stated earlier in this chapter, the Bayt al-Maswari, themselves professed Zaydis, have raised more than twenty minarets since the early 1980s for pious Zaydi clients striving for the same 'Glory, Reward, and Praise' as Imam al-Mansur. This brief survey of Zaydi minaret building in Sana'a indicates that, historically, there has not been a resolute adherence to al-Utrush's prohibition against these structures in Yemen. Rather, the subject of tower minarets was more likely a topic for theological debate, evidently with adherents championing either position. It seems more plausible that any Yemeni aversion to minarets, especially those arising later than the tenth century, were grounded in purely theological interpretations of al-Utrush's treatise rather than on any politicised opposition to the 'Abbasids' institutionalisation of the building-type (or because a local rival dynasty had accepted them). The theory of an authentically religious-based opposition is supported by such historical events as the purification of the *mihrab* in Sana'a's Great Mosque. In connection with the interdiction of plasterwork in the treatise on *hisba*, the Umayyad *mihrab* of the Great Mosque containing decoration and inscriptions was removed in the tenth century AD by order of the Qadi of Sana'a because it was thought to distract from prayers, and a subsequent prohibition was made against the ornamentation of mosques.[56] It should also be noted that, like the inconsistencies revealed in the Zaydi position on minarets, there are also historically significant Zaydi mosques, like the al-'Abbas Mosque in Asnaf-Khawlan, which display ornately decorated *mihrab*s (see plate 16).

In summary, a reconsideration of the minaret in Zaydi Yemen has elucidated two crucial points leading to a better understanding of the minaret's role in contemporary Sana'a, and for establishing a contextual framework for a study of the builders who specialise in the construction of these edifices. Firstly, the characteristic of 'height', which serves as a sign of power and affluence in an architectural context, must be contemplated as an attribute of both secular and religious building in Yemen. Like the prestige of the traditional tower house with commanding views over the cityscape, to view and be viewed, the minaret too is a marker of conspicuous consumption. Unquestionably it is also a marker of the patron's piety, a statement of the faith, and thus ultimately a symbol of Islam. Although the tower minaret is most popularly referred to in Sana'a by the term *sauma'a*, as concluded by Bloom in his etymological analysis, *menara*, derived from an Arabic word meaning sign or marker, is the most suitable word to describe this symbolic architectural element.

The second point made in the above assessment is that despite the stipulations regarding the height of minarets put forward by Imam al-Utrush of Tabaristan in his tenth-century treatise on market regulations, the Zaydi community of Yemen have historically tolerated existing tower minarets as well as made significant contributions to the proliferation of this building form. Zaydi aversions to these structures arising in Yemeni history are perhaps best understood in the context of theological debates centred on interpretations of the commands of 'Ali as conveyed in al-Utrush's treatise, and not as a politicised practice in opposition to rival dynasties and sects of Islam who have accepted the minaret as part of their religious architectural vocabulary. Tolerance of, and ultimately an embracing of, this soaring structural form as both an expression of piety and power explains the unprecedented propagation of commissions for lofty minarets in Sana'a over the past two decades.

What follows is a description of the traditional Sana'ani minaret form as premised on the one built for the al-Madrasah Mosque in the first quarter of the sixteenth century, and still built in like manner at the time of this study. Materials, structure and external ornamentation will be considered

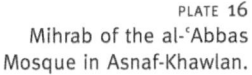

PLATE 16
Mihrab of the al-'Abbas
Mosque in Asnaf-Khawlan.

briefly before concluding this chapter with an analysis of the position of the minaret in relation to the mosque, and the significance of the prominent vertical axis. With the establishment of this historic, architectural, and symbolic framework, the concentration of my study will shift to an inquiry about the builders, and more specifically to the teaching-learning processes involved in the development of an expert craftsman: from labourer to apprentice, and from apprentice to Master Builder.

III Materials, Structure and Ornamentation

In the course of the three remaining chapters the choice of materials used, their inherent properties, and the impact they have on the minaret design and building processes of the Bayt al-Maswari will be discussed more thoroughly. What is essential here for the ensuing review of the decoration of the Sana'ani minarets is a familiarisation with the list of basic materials employed. In this overview I have included the standard English terms as well as the local Arabic terms used by the builders that I worked with in order to acquaint the reader with the technical terminologies which I will be using throughout my investigation.

The minarets of Sanaʿa, at least all those that survive, are masonry structures of either cut stone, kiln-baked brick, or a combination of both. These structures, like the tower houses, are supported on stone foundations of coarsely-cut blocks of black basalt (*hajar al-aswad*) which is impervious to damp and corrosive salts. Foundation depths extend to a solid stratum, ideally to bedrock, and in the case of the al-Maswari minarets this has required excavation to between four and six meters. The first component of the tower erected above ground is the plinth (*murabba' al-hajar*). The plinth constitutes approximately twenty-five percent of the total height of the al-Maswari minarets. In Sanaʿa, most of these bases have been built in smoothly-faced *habash* – a resilient, black, porous stone which is popularly used for dressing exterior walls on the lower stories of houses and many other buildings. In the cases of several older minarets in the city, like the al-Madrasah and the Salah al-Din (see plate 17), the square minaret bases are constructed entirely in kiln-baked brick (*ajurr*), and others, like the Ottoman al-Bakiriyyah, comprise a combination of both, with stone located on the lower portion and brick above. Plinths built in brick exhibit relief decoration which may consist of shallow niches framed by pointed or scallopped arches that may contain panels with Qurʿanic inscriptions; as well as decorative bands (*hizam manquush*) of intertwining sinuous lines; geometric grids resembling medieval Islamic magic squares which were used throughout the Islamic world as protective devices; and six-pointed stars.

The six-pointed star, resembling the star of David and once shared by Sanaʿa's former Jewish community, was known to Yemeni Muslims as the *Khatam Sulayman*, the Seal of Solomon, and is thought to have embellished Solomon's shield. Biblical King Solomon is popularly commemorated in South Arabia as being responsible for the conversion of Yemen's legendary Queen of Sheba and her subjects to monotheism, as recorded in the Holy Qurʿan. In a discussion of Sanaʿani symbolic motifs, Bonnenfant records that Solomon's Seal is the best protector of homes against evil forces and natural disasters.[57] The time of the setting sun is referred to in Sanaʿa as the 'Hour of Solomon', and during a *qat*-chewing party it is a time of meditative contemplation intensified by the echoing chorus of calls to the *maghreb* prayer.

The six-pointed star emblem is also associated with the one hundredth name of God which has not been revealed to humankind, and its 'unknownness' is associated with the empty space at the centre of the geometry. The interlocking triangles, which theoretically could be disengaged from one another, can be interpreted as the two poles of *tawhid*, or

PLATE 17
Minaret of the Salah al-Din
Mosque, Sana'a.

the Oneness of God: one is *tanzih* – God's incomparability with anything else; and the second is *tashbih* – God's similarity to everything. Murata and Chittick remark that the universe must be understood in terms of both poles simultaneously, and that this concept is encountered frequently in Islamic texts and in the everyday life of Muslims.[58] To the best of my knowledge, the Seal of Solomon is no longer used as a decorative motif, and in some cases those existing on houses have been removed or defaced, mistaken by some Yemenis as remnants of the departed Jews and now interpreted as a symbol of Israeli nationalism.[59]

The top of the plinth, whether brick or stone, is accentuated by a cornice of two or more projecting bands, and when in brick these may be individually supported on decorative dental-courses. Above the square base stands the slender columnar tower. Traditionally, the columnar portion, dating from the al-Madrasah (1519–20 AD) onward, were always constructed in brick. In the last two decades, however, several have been built in stone, and are stylistically more closely related to the

55

dominant minaret fashion in the nearby town of Dhamar, south of Sana'a, where many of the trade labourers come from. An example of the stone minarets is that of the Khalid bin al-Walid Mosque (see plate 18) erected in 1989 by the al-Maswaris. To date this was the only stone minaret which they had built, and the head masons conceded that it was considerably less labour intensive than assembling a brick tower as there were fewer individual units to handle and set. Nevertheless, it should be noted that baked brick is both lighter-in-weight and cheaper than stone, and is reputed to be more stable.[60] In discussing the evolution of Eastern Islamic minaret towers, Bloom points out that the tall thin shafts required the strength of kiln-baked brick construction in order to withstand the stresses imposed by both their own weight and by earthquakes. Stronger bricks also enabled Eastern builders to exploit their decorative potential "when they wrapped the shafts of the towers with varied and contrasting fretwork patterns".[61] Patterns on stone minarets are generally less complex by virtue of the larger size of the individual units and the hardness of the material.

The decoration of all-stone minarets, like that of stone buildings in Sana'a, consists primarily of juxtaposing polychromatic dressed stones in geometric arrangements and patterns – usually in black and beige, and occasionally other colours. If the available literary sources can be relied upon, the use of polychromatic stonework for producing a decorative effect can be dated back to at least the legendary pre-Islamic Ghumdan Palace. Basing his account on al-Razi's (ob. 1086 AD) description in *Tarikh Madinat Sana'a*, Sayyid Jamal al-Din writes that "Each of [the palace's] four facades was of a different colour of stone – one white, one green, one red, and one black".[62] Al-Azraqi's account of the construction of the later sixth-century Abyssinian Christian Cathedral in Sana'a also portrays the aesthetic use of polychromatic stone: "Between (each course of) *jurub* stones he inserted (a course of) triangular stones like a camel's hump, entering into each other, of green, red, white, yellow and black stone".[63] Bonnenfant conjectures that Cairo, Syria and Anatolia also served as possible sources for the legacy of polychrome stone decoration in Yemen. He writes that many palaces were built entirely of stone in the latter half of the twentieth century and, consequently, so too were many of the ministry buildings which were erected after the revolution.[64]

The columnar shafts of Sana'ani minarets, whether constructed in brick or in stone, are typically constituted of several superimposed, vertically arranged segments which are multi-facetted or perfectly cylindrical. The number of sides on a facetted segment is normally a multiple of four – i.e. eight, twelve or sixteen. The geometry of the square plinth below is transformed by design into a stout eight-sided base at the foot of the shaft, and this is subsequently transformed into either a sixteen-sided or a perfectly cylindrical segment of the tower. In the case of the 'Addil minaret,

PLATE 18
Minaret of the Khalid bin
al-Walid Mosque, Sanaʿa
(1989).

the slender sixteen-sided brick tower was further transformed by the masons into a cylindrical segment before reaching the underside of the calling platform (*dawwar*). Visually, the encircling platform divides the shaft into two parts, and it is supported on a system of projecting dog-toothed brick courses, each successive ring cantilevered slightly beyond the edge of the one below. The final segment of the tower rises above the platform, and terminates with an interior platform surrounded by (usually large) window-like openings. This final segment has a characteristically smaller diameter than the portion of the tower below the calling platform, and like it, the number of facets is usually a multiple of four. Sanaʿani minarets are crowned by a dome (*qubbah*) which may be either hemispherical and smooth, or ribbed (*muthallajah*) with divisions reflecting the geometry of the shaft.

The calling platform and the upper interior platform are reached by an interior stone staircase (*sullam*) which spirals around a central column (*qutub*) of roughly hewn, curve-faced stone (*hajar mudawwar*). Like the

57

tower house, the strength and integrity of the minaret's structure is vested in the combined solidity of the central masonry support and the staircase.[65] The spiral stair acts as a continuous brace, from bottom to top, between the column and the external circumference wall. Slender, cylindrical minarets with a tightly configured interior that accommodates only a staircase are architecturally categorised by Hillenbrand as one of two common minaret varieties. The other type, such as the Syrian or Maghrebi minaret, has ample interior space and often accommodates rooms (i.e. the minaret of the Great Mosque of Qairawan). Those with external stair cases, like that of Ibn Tulun in Cairo, are considered exceptional and belong to neither of the two categories.[66]

It would seem that among the early precedents for the first variety are the post-Umayyad minarets built in 'Abbasid Iraq which were prophetic of the later Saljuq minarets in Iran. Of particular note is the "so-called Manarat Mujda" of the late eighth century which is "a slender cylindrical structure of baked brick, with a diameter just large enough to accommodate a winding interior stair with sparing external geometric decoration executed in baked brick".[67] In the following centuries, the development of the "lofty, slender, cylindrical" minaret form flourished in Iran, possibly inspired by sources from the border regions of the eastern Islamic world.[68] Literary evidence confirms that these tower minarets were commonplace in towns by the tenth century, and by the eleventh their sheath of geometric brick decoration and inscriptions demonstrates a "finished assurance" that became the hallmark of the medieval Iranian minaret. Like those built several centuries later in Yemen, this "lavish decoration was not extended to the exterior of the mosques".[69] Both Lewcock and Hillenbrand cite the early use of over-all brick patterning on the relatively small surface areas of tombs and mausoleums in Central Asia as the origin for the decorative style of subsequent minarets which were built throughout Mesopotamia and Persia.[70]

The patterned brick minarets of Central Asia and Persia, and in particular the early twelfth century minarets of the Ghaznavids, are quoted by Serjeant, Lewcock, & Smith as providing the inspirational source for Yemeni brick minarets.[71] In her discussion of the Zaydi minaret at Zafar Dhi Bin, Finster explains that it "should surely be seen as having been borrowed from Anatolia or the Jazira (Northwest Mesopotamia)".[72] It is highly possible that architectural innovations and styles were transmitted from these regions to South Arabia through the oral accounts of travellers, or via drawings and illustrations on either parchment or paper. Hans-Caspar Graf von Bothmer has determined that rare colour-rendered axonometric-type illustrations (i.e. simultaneously exhibiting plan and elevation) of both the Umayyad and the Medina-type mosques were included in the parchment leafs of a hand-written copy of the Qur'an sent to Sana'a by the Caliph al-Walid in the first century of Islam,[73] thus suggesting a very early date for the transmission of architectural drawings. The Muslims' first encounter with paper was along the Silk Roads of Central Asia in the eighth century AD, where, Bloom suggests, paper was already being used in such western centers as Samarkand at least several decades prior to the coming of Islam.[74] From the thirteenth century, the manufacture of larger sheets of fine white paper in Iran, as well as a general increase in the availability and the affordability of paper, meant that artists and architects could work out designs before executing their works, and more easily transmit their designs from one place to another.[75] According to Bloom, "the most obvious new role for paper was in architectural plans", and it was by this means that the Ottomans were able to achieve an impressive degree of conformity in their building projects throughout the empire.[76]

By the time the Sanaʿani brick minaret of the al-Madrasah Mosque (first quarter of the 16th century, and the earliest surviving example of a patterned brick minaret in Sanaʿa) was erected, the circulation of designs on paper throughout Central Asia and the Middle East had become somewhat conventional practice. However, based on my studies with contemporary traditional builders in Sanaʿa, expert trade knowledge is rarely, if ever, transmitted orally, and drawings are seldom produced or referred to. None of the traditional builders I worked with or interviewed produced or used plan drawings, and only occasionally sketched small details (usually of decorative schemes) on small scraps of paper. It is possible that this was not the case several centuries ago, but as there are no extant drawings to prove otherwise, it would seem reasonable to assume that any transfer of building technology to the region would have been more competently effected via the practical works of invited, visiting, or journeying craftsmen, and adopted through the experience gained by their apprentices. This process is a less transparent one than the act of physically reproducing designs from paper (i.e. a master 'blue print'), and thus may play a role in explaining the markedly different regional stylistic developments in other regions of Yemen, such as in Sadah, Zabid, Jibla, and around the vicinity of Sanaʿa itself. This study is largely devoted to demonstrating the manner in which expertise in producing and reproducing craft within a structural framework of knowledge is achieved devoid of the type of propositional discourse expressed in either drawing or language.

The brick patterning displayed on the columnar shafts of Sanaʿani minarets, like the building materials, the structural components and the architectural form, exhibits a strong measure of continuity dating from the construction of the al-Madrasah minaret onward. In briefly discussing the adornment of several late twentieth-century minarets erected by the Bayt al-Maswari, I will refer to decorative precedents found on some of the historically prominent structures discussed so far. Sanaʿani architectural decoration, whether on houses, government buildings, or minarets, can be separated into three major classifications: geometric designs, floral motifs, and symbolic forms.[77] Qur'anic inscriptions raised in brick or inscribed in stone may be considered as a fourth category of embellishment. Inscriptions and symbols are rarely employed on the brick shaft of minarets above the level of their plinths. One of the notable exceptions is the double ring of six-pointed stars surrounding the upper portion of the Ibn al-Husayn tower (1936–37), or those on the minaret of the Qubbat al-Mahdi 'Abbas (1750–1). However, when such forms are applied in rhythmic repetition in the context of a pattern, they might be read simply as aesthetic expressions of a purely geometric nature devoid of any associated symbolic content. The al-Maswari's did not use symbols in their repertoire of motifs and the head builders insisted that there was absolutely no significance beyond aesthetic issues in any of their decoration (naqsh). Most importantly, what the decorative elements pointed to was a continuity with a past form of urban expression, recognizably Sana'ani, which in turn situated the builders firmly within a stream of tradition. Only the words of the giant brass plaque which read 'Ma Sha' Allah' (What God Wills), hung from the top of the completed plinth above the door to the 'Addil minaret, and inscriptions of 'Allah Akbar' (God is Great) commonly inscribed in the stone above their minaret doors, were considered by the builders to invoke the protective powers of God.

Like symbolic forms and inscriptions, floral motifs are also rarely applied on the brick shaft, and exceptions include the stylised tri-foliated sorghum bud (dhurah) resembling a fleur de lis. Examples of these can be found around the base of the columnar shaft of the Dawud minaret (20th c.) and on the Talha minaret (1619–20 AD) below the calling platform. For the most part, minaret decoration comprises almost exclusively geometric patterns dominated by either horizontal or vertical strings of diamonds (hirz, referring to 'an amulet'), echoed by diamond-shaped contours (diblah) or adjacent bands of chevron lines (sals). Design constraints are largely dictated by the modular quality of the thin square bricks and the limited palette of angular shapes that can be produced by chiselling them. The shaft of the first al-Maswari minaret, built for the Husayni Mosque, is perfectly columnar below the calling platform and exhibits a broad band of transforming diamond geometries raised in relief, like the sixteenth-century minarets of al-Madrasah, al-'Aqil, and the Salah al-Din. The balcony wall is decorated with two rows of small diamonds connected by a chevron line occupying the space between, mimicking the pattern on the balconies of the Musa and the Talha. Each face of the upper octagonal shaft displays a vertical string of diamonds courted on either side by three chevron lines. Similar arrangements of diamonds can be found on the upper portions of Musa, al-Mahdi 'Abbas, Dawud, and on the more recently reconstructed octagonal shaft of the al-Shahidayn (see plate 19) whose oracular tale of collapse opened this chapter.

All subsequent minarets built by the Bayt al-Maswari, including the last minaret which I constructed with them in the suburb of Madinat al-Asbahi, have employed variations on the 'diamonds & chevrons' theme (see plates 20–22). The upper portions of the towers have consistently displayed the greatest density of ornamentation, thereby constituting the visual focus of these structures and enticing the observer's gaze skyward in appreciation of their dazzling verticality. The boldness of the decoration against the planar expanses of the red-brick surfaces is accentuated by the brilliant whitewash (*guss*). This is applied to the relief surfaces of the patterns, as well as to the underside of the calling platform and the surface of the dome. White-washing is performed after the structure has been completed, and the task is executed by a distinct group of craftsmen who are specialised in this trade. It has been suggested that the brick patterning on Sana'ani minarets was not always picked out in white, and that the practice may have derived from the standard use of gypsum plaster as both a protective and decorative material on houses in the city. Lewcock, Sergeant and Smith support this claim by noting that the brick patterning of the Musa minaret (1747) was first whitewashed in 1973, and the minaret of Ibn al-Husayn, still plain at the time of their study, has since been whitewashed.[78] Early photographs, however, confirm that the relief decoration of Sana'ani minarets was picked-out in white at the beginning of the twentieth century in the same manner that they are today.[79]

In a final note on the choice of building materials, the use of traditional kiln-baked brick and stone was preferred by the head builders of the Bayt al-Maswari, like many other city builders, for a number of reasons. Firstly, it was believed that these materials are more durable than concrete block and reinforced concrete, and produce more stable structures. The builders claimed that, opposed to poured-in-place concrete structures, traditional materials could be 'controlled' – individual units and larger components could be removed, replaced and modified with relative ease and without undermining the remaining building. Thirdly, and of particular importance in the consideration of dwellings rather than minarets, clay brick and stone are widely recognised by both builders and clients as possessing more desirable thermal properties. Thick walls composed of these materials are conducive to the extreme climatic conditions of Sana'a which can experience warm and intense solar radiation in the daytime and contrastingly cool night-time temperatures.[80] Lastly, the builders consistently commented on their aesthetic preference for traditional materials over what they referred to as 'modern' or 'Western' materials. This preference was largely based on the assumption that decorative stone and brick facades have historically distinguished the urban landscape of their city, and these contemporary builders felt themselves to be directly involved in a practice that perpetuated a noble craft and a product which retained its prestige in the eyes of the public.

PLATE 19
Minaret of the al-Shahidayn Mosque,
Sanaʿa.

PLATES 20–22
Examples of minarets built by the
Bayt al-Maswari in Sanaʿa 1980–1997.

PLATE 20
Al-Husayni minaret.

PLATE 21
Zuheiry minaret.

PLATE 22
Bileili minaret.

IV Position and Axes

With the Abyssinian conquest of Yemen completed, the Axumite General Abrahah seized power in 537 AD and undertook the construction of the largest Christian edifice south of the Mediterranean: a pilgrimage cathedral dedicated to the legendary place where Jesus came to pray during his years in the wilderness. Abrahah's reign was brief. During his campaigns against Mecca, his army was forced to retreat from a 'plague of flies' (which historians believe was an outbreak of smallpox), and Axumite rule was soon after relinquished to the conquering Sassanians in 575 AD.[81]

After the Abyssinian Christian Cathedral of Sana'a, or *al-Qalis* as it is referred to, was destroyed in the late eighth century AD under the command of the appointed Muslim governor, al-'Abbas b. al-Rabi', a pit approximately two meters in depth was left in the most eastern quarter of the old city where the martyrion of the cathedral once stood. Lewcock reports that the foundations, or lower walls, are visible, lining the pit which is fourteen meters across.[82] One Yemeni acquaintance suggested to me that the still extant hole in the ground symbolically represented the inverse of the Ka'bah at Mecca which the Christian Viceroy of Sana'a set out to destroy with his army of soldiers and elephants in the year 570 AD – known as the Year of the Elephant (*Am al-Fiel*) and the year that the Prophet Mohammad was born. Therefore, for him, the hole signified the defeat of Christianity in the face of Islam in South Arabia. What I found to be most provocative in his symbolic analogy was the suggestion of a strong vertical axis represented by the inverted relation between the amorphous hole left by the cathedral and the positive, axial volume of the Ka'bah. In the local context, it was the nearby minarets in the south-eastern quarters of the old city which struck me as being most apparently in juxtaposition to the pit. The minaret, more so than any other architectural form in Islam, embodies the importance of the vertical axis, reaching heavenward and, as I have shown, serving as the ultimate outward symbol of mankind's patronage to his Faith.

Historically, and in the case of all those built by the al-Maswaris in the latter quarter of this century, the minarets of Sana'a stand as independent structures from their respective mosques, and often at a considerable distance from the prayer hall. This, along with the fact that they have remained stylistically consistent in the face of changing tastes for mosque architecture (compare the typical Zaydi cubical mosque with the domed Ottoman-style mosques), indicates that many were likely commissioned independently and at later dates, and that they have been traditionally built by teams of craftsmen with a specialised expertise in the trade. The square bases of Sana'ani minarets are predominantly oriented along one of two major horizontal axes: either in line with the *qibla* wall and oriented with one face toward Mecca; or, alternatively, aligned with the cardinal axes.

The Master Builder's decisions regarding minaret positioning at the sites I worked on were guided by considerations of available space, slope of the land,[83] soil conditions, and the recognised aim of both builder and patron to maximise the visibility of the structure. The importance of the vertical axis superseded that of the horizontal, unlike that of the mosque which obeys "an invisible set of lines of force which attract all points in the periphery toward the Centre [Ka'bah]."[84] As noted by Robert Hillenbrand in addressing the issue of minaret positioning in Islamic Cairo, many were "unusual or unprecedented" indicating that the minaret was now "valued less for its actual, or symbolic religious function and more for its role as a marker or articulating feature".[85]

In Nasr's spatial analysis of a typical domed mosque building, the symbolic relations which he draws between the basic architectural components implies an articulated vertical hierarchy between the heavens and earth. This abstraction could be transposed onto a symbolic understanding of the domed Sana'ani minaret tower with its square stone plinth and octagonal shaft as it has been described in this chapter. Nasr's theologically-biased evaluation both contrasts with, and perhaps compliments, the more strongly secular perception of the minaret thus far derived. His reference to minarets in the final sentence of the passage also symbolically relates the vertical expression of these structures to power ("majesty"), even if it is qualified here as Divine:

> The dome, while creating a ceiling which protects from both heat and cold, is also the symbol of the heavenly vault and its centre the axis *mundi* which relates all levels of cosmic existence to the One [God]. The octagonal base of the dome symbolises the throne and pedestal and also the angelic world, the square or rectangular base the corporeal world on the earth ... The external form of the dome symbolises the aspect of Divine Beauty, or *jamal*, and the vertical minaret Divine Majesty, or *jalal*.[86]

Interestingly, and as will be described in greater detail in the related sections in this study on the art of building, the axial centre of the minarets built by the Bayt al-Maswari is embodied by a metal rod (*qasabah*) which extends from the bottom of the structure and pierces the apex of the dome to support the bronze crescent moon (*hilal*). During construction, all measurements were radially marked on the horizontal plane by a rope fixed to this pole, and the pole was vertically extended with additional threaded segments as the minaret structure heightened around it. The metal pole was anchored in the foundations, and was intricately bound to the physicality of the entire structure by way of measurement: from the circumference of the stone column that embraced it; to the wind of the staircase; to the inner and outer geometries of the tower walls. Only at the level of the cupola, which completed the primarily vertical design, was the radial rope employed to calculate construction points on both the horizontal and the vertical planes of the dome's curvature. Beyond the dome the axial pole no longer served to mark extensions in the horizontal plane, but functioned exclusively to support the affixed Islamic symbol of the crescent moon (*hilal*) high above the city.

The nature of the minaret's dominant vertical axis and subordinate horizontal dimensions was reflected in the hierarchical ordering of the work team and the physical placement of the builders in accordance to their individual status on the site. There were essentially three major divisions amongst the traditional builders which reflected the levels of trade experience, expertise, and assigned responsibility. These were, first, the level of being employed as a common labourer; secondly, the stage of being taken on as an apprentice; and lastly, achieving the status of a Master Builder, or *usta*. Within these levels were further hierarchical divisions cultivated and refined by social and political struggles amongst the builders, as well as by hereditary status and levels of education. The al-Maswaris built their minarets from the inside-out without the use of scaffolding, therefore the staircase was extended in conjunction with the level of the walls and central column, and served as the principal means of access to the elevated work site. The delivery, unloading, and processing of construction materials were executed at the ground level by several labourers. Others lined the interior winding staircase, passing on materials from the bottom to the top, and discarded materials and debris from the top to the bottom of the vertically extending structure. At the top of the processing chain was the head labourer, or apprentice, whose chief task was to organise arriving materials, prepare them, and pass them on to the Master Builders as they were required. As the title implies, the Master Builders did the actual building which consisted mainly of measuring out geometries, setting courses of masonry, and verifying the vertical and horizontal levelness of the erected surfaces, as well as overseeing and regulating the conduct and performance of the work team below. The Master Builders, who commanded both the top of the minaret's vertical structure and the upper echelons of their professional hierarchy, were also responsible for establishing client relations and contracts, ordering materials, and administering finances and various other aspects of their construction projects.

In the following chapters I will address each of these three levels of the building trade in turn, beginning with an analysis of the disciplinary training experienced by the labourers, followed by a study of the apprentices' acquisition of trade-related skills and knowledge, and concluding with a discussion of the continuing education and the expertise of the Master Builders. Each chapter commences with a discussion of the construction techniques and practices employed at each successive phase of

66

minaret building, starting with the foundation and base, continuing with the brick shaft, and finishing with the construction of the dome. I have metaphorically related these building phases with the stages in a builder's career. My study has relied almost exclusively upon the experience I gained during the course of the year which I laboured for the Bayt al-Maswari. Although my research has focussed on the lives, the work, and the training of this particular family and their team of employees, I became sufficiently familiar with other teams of builders in the city, either through brief periods of work experience or regular visits and interviews, to present my findings and analyses in this study as being representative of the teaching-learning processes of traditional Sana'ani builders.

Chapter Two

FOUNDATIONS

Training Labourers in a
Traditional Apprenticeship System

To interiorise life itself and to become aware of the inward dimension, man must have recourse to rites whose very nature is to cast a sacred form upon the waves of the ocean of multiplicity in order to save man and bring him back to the shores of Unity.[1]

Seyyed Hossein Nasr

Studies on Yemeni architecture have overlooked the manner in which skilled spatial judgment and technical expertise are inculcated in the individual builder. In the recent past, all young Yemeni men seeking a career in the building trade negotiated an apprenticeship with a Master Builder, or *usta*, in an urban centre. As recently as 1970 there were no professionally trained architects in Sana'a, and all building and architectural decoration were the exclusive domain of these Masters. Although this situation has changed considerably with the introduction of modern building technologies, the *usta*, as a socially constituted agent, continues to manipulate significant power in the production and reproduction of both his trade and the built environment which, in turn, embody the complexities of contemporary Yemeni culture and education within an Islamic context. Before addressing the issues of knowledge and training, it is important to first briefly contextualise the social status and profession of the builders within the larger Sana'ani framework. It should be noted that the system of social ranking has changed considerably since the revolution of 1962, and continues to evolve.

Social Rank and Builders

Many anthropological studies of Yemen address the country's class structure(s) in great detail, including for instance those of Dresch, vom Bruck, Mundy, Weir, Meneley, Stevenson, and Gerholm, all of which are concentrated on regions included in the former territories of the Yemen Arab Republic.[2] These studies have largely focused on the politics and commerce of the tribesmen and upper classes, amongst whom issues of social ranking were prominently discussed and played an overt role in identity construction. I would contend that the traditional social divisions based on birthright rank are more strongly enforced and acutely felt in the rural towns and villages of the country than in the larger urban centers like Sana'a. In the contemporary urban sprawl of Sana'a, new social divisions coexist with the surviving remnants of the old order, but they are tangibly carved along newly emerging economic lines, reflected in the physical segregation between the rich and the poor of the modern city districts. This contrasts sharply with the past when urban quarters of the walled city were not rigidly divided according to differential status, and wealth and status did not necessarily coincide.[3]

Traditionally, Sana'ani urban society was a strongly class(caste)-based one, stratified in descending order of the *sada* (sing. *sayyid*), the *quda* (sing. *qadi*), the *manasib* (sing. *mansib*), the *Bani Khumis* (Lewcock refers to this class as the *muzayyin*),[4] and the *akhdam*.[5] Certainly prior to the revolution, stratification of the highland Zaydi population was all-pervasive: "In sumptuary regulations, in historical and legal writing and in the oral culture of the religious elites of the Imamate", writes Mundy, "we find a vision of a social order as a hierarchy of ranks".[6] The elite status of the *sada* has been historically vested in their alleged bloodline connection to the Prophet through Fatima and 'Ali, while that of the *quda* is prefaced on their descent from an original class of judges. Though, as vom Bruck points out, the *sada*, by virtue of their exclusive rights to supreme leadership, "became the main targets of the revolution",[7] they, along with the *quda*, continue to have a somewhat decisive influence on the cultural life of the present.[8] In face of growing demands from Sunni-oriented political groups for an Islamic state, many of the Zaydi *'ulama* have become more strategic in the manipulation of their social identity in order to ensure the survival of their class. In her study on this issue, vom Bruck notes that "most have attempted to maintain their Zaydi identity whilst seeking alternative ways of accommodating a political reality which challenges the notion of descent-based authority as enshrined in Zaydi-Hadawi doctrine".[9]

Next in the Sana'ani line of rank, the middle group of the *manasib* (literally 'dignity, rank or position') consists of builders or masons, and other tradesmen such as "gold, silver, copper smiths and black smiths; plumbers; carpenters and turners; dyers; men who burnish blades and work dagger hilts, make hooka tubes and mattresses; painters of inscriptions; and porters".[10] These pursuits are considered to be 'honourable', and those people (also referred to as *arab*) engaged in them are understood to have a 'tribal ancestry'.[11] Weir notes that although in most of the former Yemen Arab Republic, trades such as black-smithing and carpentry were considered demeaning for the higher classes, in some instances, such as the western highland town where she conducted her research, they were not conceived as inferior occupations and some *sayyids* engaged in them as their primary source of income.[12] Members of the *manasib* class (or the in-between classes in Mundy's tri-partite sense) are related by blood with the free highland tribes, and, as differentiated from the men of religion (the *sada* and the *quda*), they are recognised as the men of the sword and the plough.[13] Historically it was this group of able-bodied citizens who were called upon to defend Sana'a when it came under siege.[14] The *Bani Khumis* ('Sons of the Fifth')[15] (or *muzayyin*) and the *akhdam* comprise the lowest levels of the social order, engaged in 'despised' trades and providing services which the other three groups have traditionally refused to participate in. The former group consists of "cobblers; makers of leather sheaths; brass founders; saddlers; tanners; brickmakers; barbers; bath attendants; cuppers; coffee-house proprietors; butchers; and farmhands whose job it is to look after the municipal gardens", and the latter, the most despised group, are largely street sweepers.[16]

A softening of the rigid class structure was already apparent in the decade following the revolution. The institutions which evolved with the new centralised state, importantly including the heavily Egyptian-influenced educational system, in conjunction with the rapid economic growth which accompanied the mass migration of male labour to the oil-rich states of the peninsula, together entailed the "gradual demise of the status differences cultivated under the old regime".[17] It became increasingly apparent in the 1970s that "the market (and marketable labour) was set to be the dominant institution in the new Yemen".[18] "In most of Yemen market-trading, including qat-trading, was traditionally despised by 'tribesmen' and sayyids", remarks Weir, "but became more 'respectable' during the 1970s" in light of the financial rewards. Though butchery has remained a despised occupation – and this despite the burgeoning wealth of that occupation's members – a café was nevertheless opened by a tribesman (*qabili*) of high birth in the highland town of Weir's study.[19] The fact that such traditional boundaries have been increasingly transgressed in the past decades points to the rising influence of market conditions and financial wealth in the country's post-revolutionary social structure. The state army, government ministries, higher education and new job opportunities responding to Yemen's immersion in the global economies, have ruptured the once direct correlation between birth-right status and occupation. This challenge to the old order is most poignantly felt in the larger urban centers where resources are concentrated, links with global markets and opportunities are more apparent, and individual manipulation of possibilities and alternatives has greater probability of being realised.

The builders whom I worked with never spoke of social rank, though clearly, based on the traditional models of hierarchy, the family of Master Builders would be classified as *manasib*. The now-deceased father of the oldest generation of the building family came to Sana'a in the 1930s from

Maswar, a highland region west of the capital. 'Ali Said al-Maswari settled in the western suburb of Bawniyyah where the majority of the family members have remained until the present day. Knowledge of the family genealogy prior to 'Ali Said is sparse. Unlike many of the *sada* and *quda* who have cultivated elaborate genealogies tracing their origins through individuals back to the founders of the Faith, typically members of this lower social class trace their origins to "a particular place where [their] patronymic house is known for its relation to the land, its honourable occupation and its marriage alliances".[20] The branch of the Bayt al-Maswari whom I worked with (the direct descendants of *Usta* 'Ali Said)[21] constructed their identity primarily around their engagement in the building trade, and the reputation the family had earned region-wide as excellent craftsmen. A steady stream of building contracts for prestigious projects (which included, in addition to mosque minarets, houses for the family relations of President 'Ali 'Abdullah Salih) have secured a comfortable middle class position for this family of builders in the contemporary Yemeni economy. This has provided some younger members of the *bayt* with the opportunity to pursue a formal education and alternative vocations. The form of education that these young men[22] receive in the classroom is markedly different from the training they would have received on the building site, and it is the latter, which inculcates both trade knowledge and a disciplined sense of moral being, which will be the focus of the ensuing study.

Knowledge, Guilds and Training

Tapper & Tapper importantly note that:

> The English term 'knowledge' is ambiguous, covering two kinds of 'knowledge' that in most other languages (and philosophies) are kept terminologically and conceptually quite distinct: one is theoretical, scientific, analytical, learned from texts by memorization; the other is personal, practical and holistic, and comes from experience and intuition.[23]

Their discussion emphasises the distinction made between the two in the Islamic world by the Arabic terms *'ilm* and *ma'arifa*, both of which translate in English to 'knowledge'. *'Ilm* refers more specifically to the extensive but limited body of religious science of the *'ulema*, which is appropriated through memorisation and recitation of the religious texts. It also includes knowledge of the other sciences and that which can be known 'by the book'. *Ma'arifa* is regarded as "the esoteric, personal knowledge gained from experience", and is often extended to include such non-

71

religious knowledge as craft skills. The relation of craft skills in the Middle East with this esoteric type of knowledge may be explained by the fact that, in many Islamic towns and cities, the craft guilds were "organised in direct association with or imitation of Sufi orders". For a Sufi, goals of personal enlightenment (*ma'arifa*) are achieved by following the Path (*tariqa*) towards Truth (*haqiqa*).[24]

The Sufi orders, though existent in Yemen since the early Islamic period, never found fertile ground in the highland Zaydi regions of the country, and have historically existed there in small numbers compared to the predominantly Shafi'i regions. Buchman remarks that Sufi aspirations to attain *ma'arifa* directly in one's heart are illogical to the Mu'tazilism theology of the Zaydis, which maintains that "God is totally incomparable (*tanzih*) with the world and would argue that the human rationality faculty (*'aql*) is the only reliable means of obtaining knowledge from divine revelations".[25] Craft guilds in Sana'a, however, have traditionally played an important role in production, trade, and the regulation of the market place, but seemingly not for builders. None of the builders that I worked with or interviewed during the course of my study were affiliated with a guild, and claimed that they never had been. Usta Sarhan ar-Rawdhi, the member of a prominent family of Master Builder for many decades, claims that, to his knowledge, there has never been a *niqabah* (guild) for the builders of Sana'a.[26] Sometime in the 1960s,[27] according to ar-Rawdhi, there was an unsuccessful attempt to establish an association of builders which was meant to serve principally as a charity for *usta*s, but was not meant to examine the competence of the craftsmen, as described by Ibish.[28] In fact, there is no '*Shaykh*' of the builders, but rather building remains a vocation open to enterprising individuals, and the quality of the work and the success of young builders in becoming Masters of their trades, is entirely in the hands of the individual *usta*s.[29]

The absence of a builder's guild in Sana'a is not surprising considering that many of the Master Builders, as well as their teams of labourers, reside in villages and towns outside the capital, and only venture to the city when they receive commissions for work.[30] There is also no central point in the city, or in the suq (such as is the case for many craftsmen), where the builders congregate by necessity or for association. The nature of their work requires that they are dispersed throughout the city (or region) and not necessarily in (regular) communication with one another. For many, their work is seasonal, balancing their professional involvement in building with agricultural pursuits in their village. In the absence of a controlling body which regulates the quality of construction, use of materials and aesthetic considerations, it is the *usta*'s obligation to ensure that the necessary knowledge is passes along to his young team of builders in the interest of safeguarding the integrity of both the craft and his own *bayt*.

In my study, I consider both the technical knowledge, or *'ilm*, and the personal knowledge, or *ma'arifa*, which the builders receive. The link between the individual's education and training as a Muslim, a craftsman, and as a member of a responsible community of 'experts', will provide an important key to understanding how spatial knowledge and expertise is inculcated in a traditional Sana'ani builder. Traditional builders have been defined for the purpose of this study as those employing indigenous materials and construction methods, and deriving their expert knowledge through apprenticeship as opposed to a technical or formalised education process. During the course of my fieldwork I apprenticed myself to a reputed family of Master Builders, the Bayt al-Maswari, who have specialised since the early 1980s in the construction of minarets and who have been largely responsible for the renaissance of Sana'ani-style minarets that pierce the skyline of the newer city quarters. Apprenticeship as a methodology has enabled me, as ethnographer, to merge the subject of study with the object of study, and most importantly to *learn about learning* in a context in which formal technical training, engineers, and drawn plans are non-existent.

During the course of my labouring with the builders I perceived a correlation between the hierarchical stages of becoming a Master Builder and the theological concepts of *islam*, *iman*, and *ihsan*. Seated in the *mafraj* on the fifth storey of my shared house in the old city, I regularly peered down onto the nearby primary school. The bird's eye view of activities always provided tremendous amusement, and more than once I witnessed young boys escaping from the school compound along classroom rooftops, and shimmying down the boughs of a tree to safety on the other side. At the start of both morning and afternoon sessions, all the children assembled in files in the school yard to perform exercise drills. Clearly, these drills were not designed to challenge and exhaust the rambunctious energy levels in children, but served rather as disciplinary exercises. Even when not watching from my fifth storey perch, I heard them through the closed shutters of my room, counting in unison "*wahid, ithnayn, thalatha, arba'a*", to the rhythm of the drills. Afterwards, a different student each day, usually from a more senior year, recited Qur'an while the hundreds of other children were meant to stand at attention, listening to the broadcast of the Holy passages over the crackling loudspeaker. I periodically took my observations and questions about Yemeni education to my fellow builders at the minaret site and inquired about their own schooldays. Their accounts, particularly those regarding religious education, led me to make comparisons between the stages of Islamic training and their training as builders, which they too found familiar.

The Qur'an speaks of *islam* (submission), *iman* (faith), and *ihsan* (virtue). These three dimensions of the religion are defined more clearly in the Hadith:

Abu Huraira said, The Prophet, peace and blessings of Allah be upon him, was one day sitting outside among the people when a man came to him and asked, 'What is *iman* (faith)?'. He said: '*Iman* is that thou believe in Allah and His angels and in meeting with Him and in His messengers and that thou believe in being raised to life (after death)'. He asked, 'What is *islam* (abandonment to the Divine Will)?'. The Prophet said: '*Islam* is that thou shalt worship Allah and not associate aught with Him and that thou keep up prayer and pay the *zakat* as ordained and fast in Ramadan'. He asked, 'What is *ihsan* (spiritual beauty)?'. The Prophet said: 'That thou shall worship Allah as if thou seest Him; for if thou dost not see Him, surely He sees thee'.[31]

Briefly, the concept of *islam*, or submission, stresses the embodiment of practice: as children learn to recite Qur'an without necessarily understanding, young labourers and apprentices are likewise trained to perform tasks deprived of explanation. Once the apprentice is chosen to work closely alongside the *usta* he gains some understanding of the processes through careful observation, faithfully replicating the trade skills inculcated through the teachings of the Master. His final graduation to *usta* marks the transition from a state of faith and understanding, or *iman*, to one of intentionality, or *ihsan*, at which point his thoughts and actions are absorbed with the drive to reproduce the 'beauty' of the craft over which he has acquired mastery.[32] As opposed to my own architectural training which first instilled a broad knowledge base for the manipulation of abstract spatial concepts on drawn plans before proceeding towards an expertise of 'the particular' in the construction process, the Sana'ani apprenticeship begins by mastering the particular, or the actual 'making', as embodied in the practices he is taught to perform, and moves slowly toward a global understanding of the complex spatial relations wrought by the trade 'intention'.

In the remainder of this chapter I will focus primarily on the first stage in the apprenticeship system, labouring, and the inculcation of discipline and practices. I have drawn upon the performative characteristic of *islam* and the early education of a Muslim to more fully understand the processes of formation of a Sana'ani labourer. Although *islam* is often used interchangeably with *iman*, as both spring from faith, several hadith clarify the difference between the two whereby "*iman* strictly indicates the acceptance of a principle which is the basis of an action – the theoretical side – and *islam* the action itself – the practical side of man's life".[33] Both the training of a labourer and that of a child in performing rites and rituals are concerned with life's practices, and therefore more closely associated with *islam*. The objective of the training in both disciplines is the sedimentation of a specialised bodily knowledge which becomes manifest in actions, and fine-tuned to respond reflexively to a given context or situation. The deeper significance of these actions, as an expression of knowledge (both *'ilm* and *ma'arifa*), will only be mastered at a later phase in the learning process by the vocationally matured and motivated individual.

The chapter is divided into two sections: the first, and shorter of the two, aims to give a brief description of the processes and considerations involved in the initial stages of erecting a minaret. A connection is made between laying the foundations for the structure to be raised above and the early education and inculcation of basic practices in initiates. The following section discusses religious training in Yemen's formal school curriculum, and the training of labourers in the building trade. I drawn upon the concept of *islam* in order to demonstrate the similarities between the two training systems, and to expand the contextual framework within which the learning processes of a Yemeni builder are motivated. I will also draw upon the works of Bourdieu and Bloch to advance an understanding of this largely non-linguistic, embodied knowledge, and, importantly, the manner in which it is inculcated and becomes habitual and efficient. Throughout the book I will draw out the microphysics of the minaret building site – in detailed terms of the building programme; site conditions; the relations between builders, between builders and other trades, between builders and clients, and between builders and the larger Sana'ani social, cultural and economic context – as a place of work and learning. By doing so, I aim to move beyond Bourdieu and provide a more detailed and satisfying account of the manner in which structured knowledge becomes gradually inculcated in the agent.[34]

I About Building: foundations and principles of structure

Setting the foundations and defining the principles of structure is the first and most crucial step in erecting any edifice. These two factors determine the expressive and aesthetic potentials of the building, as well as its structural integrity, both during the course of construction and over the period of time which the structure is used. Principles of structure are not merely a response to the universal laws of physics, but, as evidenced by the immense variety of architectural styles world-wide, are enriched by a number of locale-specific considerations including the nature of available building materials and construction methods, and, very importantly, by the traditions and decisions imparted by the architect-builder. The Master Builder's (*usta*, plural *asatiyyah*) contributions in terms of both habitual and innovative decision-making are, like the expressive qualities of the edifice, governed by a set of principles. These are grounded in the cognitive capacities and limitations of his human mind, and conditioned by the socialising processes of both his particular profession and by the value systems of his society. The weaving of the two socialising processes, that of the trade and that of society, supplies a superstructure to the individual's decision-making which expresses a trade expertise and professional credibility. This complex union of technical skill and cultural sensitivity must be mastered over the course of the apprenticeship by the young builder in order that he may be set firmly upon the path toward professional enlightenment and achieve the integrity of a Master.

The siting of a new minaret commissioned for the 'Addil Mosque in the former Turkish quarter of Bir al-'Azab, west of old Sana'a, was, like all previous projects, selected strategically by Muhammad al-Maswari, the chief Master Builder and head of the extended family. His processes of site selection took into account several factors including the existing soil conditions, the slope of the land, the availability of space in the vicinity of the mosque, and, most importantly, he aimed to maximise the visibility of the tower. It was understood that the minaret would act as an important marker signifying the presence of the religious institution within the neighbourhood, as well as the heart of that particular neighbourhood in relation to the rest of the city. As discussed in the previous chapter, despite the more prevalent Sana'ani use of the word *sauma'a* in reference to minarets, Bloom's theory regarding the term 'minaret' as being "ultimately derived from an Arabic word meaning sign or marker", seem a more accurate reflection of the builder and patron's emphasis on verticality.[35] In the case of the 'Addil minaret, a site was chosen in the mosque's forecourt and adjacent the street where it would be prominently viewed from a number of directions.[36]

Next, the ground was broken and excavated to a depth prescribed by an engineer based on soil inspections and considerations for the intended height of the tower. I was initially told by the *asatiyyah* that they did not use the services of an engineer, but rather they made all necessary structural decisions on site based upon experience and their inherent trade knowledge. The latter assertion was usually gesticulated by tapping their temples with their index fingers. I later encountered an engineer at the site who claimed to be a family friend of the Bayt al-Maswari and a consultant for their projects. When I queried Ahmad al-Maswari, Muhammad's younger brother and next-in-command, about the role of the engineer, he denied any professional associations with him and reasserted claims to the family autonomy over all structural decisions. For some time, this was a source of minor confusion for me. The same engineer came to the site on several occasions during the course of the year and was always received amicably and with great respect. He once told me that he had completed his undergraduate studies in engineering in China and received a Masters degree in London. He insisted that it is he who determines the depth of the minaret foundations for each site and the limits on how high they can raise the structures. After several months the head *usta* confirmed to me, almost reluctantly, that some structural decisions were indeed made by this engineer. These traditional builders were fully cognisant of the myths that enshrined their trade knowledge and which delineated their expertise as distinct from that of the modern contractors in the eyes and minds of their Yemeni clients. I sensed that admissions to the employment of engineers and modern technology conceded a share of the power over the expert discourse to the 'other side'.

The orientation of the four faces of the foundation walls may be aligned along the axis and cross-axis of the *qibla* direction,[37] which is slightly north-west in Sana'a, or, alternatively, in conjunction with the cardinal directions thereby aligning approximately with the rising and setting of the sun. It has been suggested by Lewcock, Serjeant & Smith that the true north orientation of the sixteenth-century minaret of the al-Madrasah Mosque was possibly set to correspond with its use for astronomical or chronometric purposes.[38] The Bayt al-Maswari builders have subscribed to the latter positioning for most of the twenty-five or so minarets they have built in and around Sana'a since 1980 and the time of my study, including that of the 'Addil minaret (see figure 1). However, despite the extreme precision of their geometric constructions, there was absolutely no evidence to suggest that these new minarets served any astronomical purposes.

The excavated hole, five to six meters deep on average and reaching bedrock, was subsequently filled in with large pieces of rubble and concrete, behaving like a large foundation footing for the stone base and brick tower which would be built above it. The rubble used in the foundation is *ju'm*, a hard black basalt impermeable to damp and salts which my rise by capillary action through the base of the edifice and corrode its structural integrity. It has been suggested by other authors that the term *mawthar* is also used to denote this basalt stone,[39] though I did not encounter this term on the building site. This same method of constructing the foundation has been

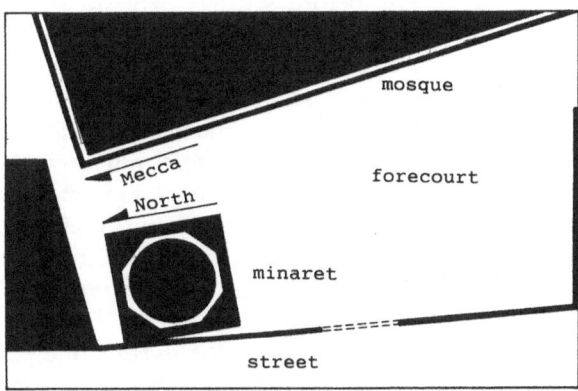

FIGURE 1 Figure plan of the 'Addil mosque and minaret and their respective orientations.

employed in Sana'a possibly since antiquity as indicated by archaeological investigations, and is mentioned by Jamal al-Din in his mid-eighteenth-century description of the structural organisation of Sana'ani tower houses in *City of Divine & Earthly Joys*.[40]

In some cases, to initiate the construction processes, an owner, more often of a house, may sacrifice an animal, such as a goat or a chicken, and will baptise the foundation corner stones with blood to protect his future abode against ill fortune and the evil eye. More commonly home owners in Sana'a will place the horns of a bull, ibex, gazelle, or goat on the exterior corners of their dwellings,[41] or include Qur'anic verses in the exterior plaster decoration, for protection against malevolent *djinn*, the evil eye, and lightening. To the best of my knowledge, the sanctioning of the building with blood or charms was not enacted during the erection of new minarets or other buildings associated with religious institutions, and was certainly not practised by the Bayt al-Maswari. However, a plaque reading *"Ma Sha' Allah"* was placed above the doorway of the minaret tower, and several workers did carry their own protective devices against evil *djinn* on their bodies.

The first several courses of stone resting directly upon the footing and extending slightly above grade are of a hard dense black basalt, *hajr aswad (al-hajr al-asawda' or sawda')*[42], which is resistant to damp rise and corrosive salts. The stone mason, or *muwaqqis*, cut these facing stones to size using a heavy iron hammer, or *zubrah*. He verified their dimensions with a pre-cut steel reinforcing bar, and their squareness with a ninety degree angle, or *zawiyyah*. The face of each unit was left rough and unfinished, and each was cut to a rough pyramidal-shaped wedge on the backside in order to achieve a stronger and more penetrating bond between the facing elements and the infill of the base. If necessary, the *usta* would chisel the blocks to a more exact height and width with his adze-like hammer, *mitraqah* or *mifras,* before setting them in quicklime mortar. Above the final course of basalt the *usta* began setting the more perfectly dimensioned and smooth-faced blocks of black *habash* stone. *Habash* stone is a hard, slightly porous volcanic stone quarried south of Sana'a in the region of Dhamar. Prior to the 1980s the stone was painstakingly cut to size by hand and dressed on site by the *muwaqqis*, but now the stones are machine-cut and delivered with need for only minor modifications. The *muwaqqis* bush-hammers a dense pattern of fine diagonal hash lines, *raqamah*, in neat rows across the face of each stone in order to give the finished surface of the minaret's black stone base, or *murabba'a al-hajar*, a mildly rusticated, but controlled, aesthetic. The word *raqamah* refers to the incision patterns made on the face of the *al-habash* stone, and is derived from the verb *raqama* meaning appropriately 'to brand, imprint, or to stripe' (see figure 2).[43] With reference to the Arabic names for stone, those commonly used in Yemen to describe black basalt, *hajar aswad*, and black

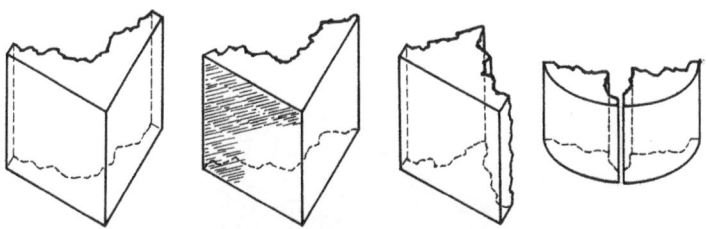

FIGURE 2 Stone cuts: a) corner *habash* stone; b) corner *habash* stone with hammered line pattern; c) facing *habash* stone; d) rough cut stones for column.

volcanic stone, *hajar habash*, both make reference to the colour of the stone: the translation of *aswad* is black, and *al-habash* refers to Abyssinia or Ethiopia and thus, in the case of the stone name, refers to the 'black' people of this neighbouring land across the Red Sea who once upon a time ruled Yemen. Colour, or alternatively place of origin, are salient characteristics commonly elected by the local building industry for categorising different stone types.

In the initial stages of building a minaret, the stone courses making up the exterior faces of the foundation above grade must be made level in a systematic and rigorous manner. Any compromise to the perfect verticality of the structure will result in its failure, succumbing to the forces of gravity under its own unevenly distributed weight. Verifying the levelness was achieved with a simple hydraulic system consisting of a long and narrow, clear plastic tube partially filled with water. Once the *habash* stones had been positioned at each corner of the square-planned base, the *asatiyyah*, or Master Builders, verified that their vertical elevations were equal to one another. The senior *usta* held his end of the plastic tube upward at one corner, and his partner did likewise at a second corner, counter-clockwise in relation to the first, leaving the slag tube resting on the ground between them. Either end of the tube heights were adjusted in a dialogical fashion until it was determined that the water levels, evidently marking the same vertical elevation in space, corresponded to the top heights of both corner stones (see figure 3). This procedure was repeated around all four faces of the base, proceeding in the counter-clockwise direction, using the first stone as a benchmark for all subsequent measures and making any necessary adjustments to the settings of the individual corner stones. Following the verification of the first five or so courses in this manner, subsequent courses were carefully levelled on the vertical plane using a plumb-line, *mizan*, which was lowered down along the face of the exterior walls from above.

The plan of the stone bases (see figure 4), or *murabba'a al-hajar*, built by the al-Maswaris are standardly 4.5 meters square, however the height of the

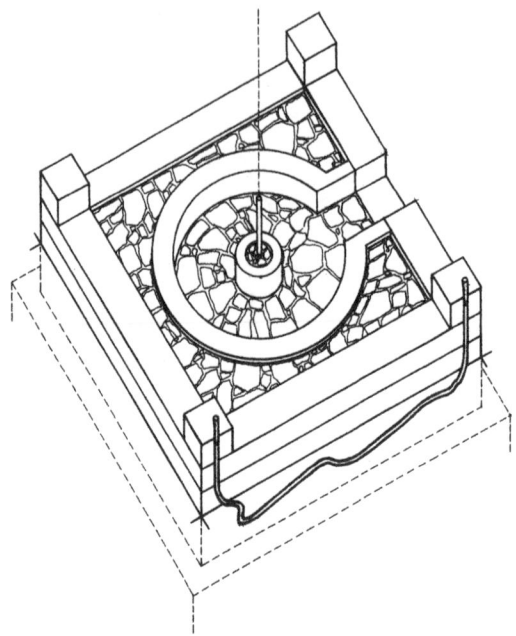

FIGURE 3 Axonometric view of minaret base with hydraulic
system for determining elevations of corner stones.

base tends to be varied in proportion to the height of the minaret, and
constitutes roughly one quarter of the structure's total vertical measure. The
stone base of the 'Addil minaret reached fifteen meters and was constructed
with inner and outer walls of faced stone and infilled with rubble masonry.
During this early phase of construction, a team of twelve builders were co-
ordinated, each with individual tasks. Two Master Builders assisted by an
apprentice set the exterior courses of cut stone which had been prepared by
the stone mason; meanwhile the *thana'a*, or second in command below the
Master Builders, oversaw the infilling of the base with the direct assistance
of one labourer, and he also supervised the remainder of the labour corps.
The corps included one labourer mixing the gravel mortar, three who
systematically transported buckets-full to the base, one who carried large
uncut stones, and another who collected smaller bits of rubble in order to
complete pockets of the infill.

Once the base has risen several stone courses above grade, the circular
interior space of the spiral-stairwell was defined. This circulation shaft,
roughly 2.4 meters in diameter, was centrally positioned within the plan, its
empty space and spiralling stairs encompassing the solidity of a central
masonry column (see figure 5). The convex-curved stones for the column's
exterior face were cut by the stone mason guided by a bent steel bar which

FIGURE 4 Plan of the minaret base.

FIGURE 5 Section drawing through the minaret base.

served as a template for the column's fixed diameter of approximately sixty-five centimetres. The column was infilled with mortar and rubble masonry, and within this, at the axial core, was a metal rod which rose through the entire height of the column, ultimately piercing the dome of the minaret some fifty-five meters above the foundations.

Once the stone base had been completed, a large brass plaque crafted in the old city *suq* and reading '*Ma Sha' Allah*', 'What God Wills', was hung from a chain over the doorway into the minaret, offering God's protection to all those who enter the work site. When used verbally, '*Ma Sha' Allah*' expressed surprise or bewilderment, and was also commonly uttered rhythmically by men when climbing the staircase in a Sana'ani tower house to warn unwary female occupants of their arrival. Visitors to the minaret site repeated the phrase while climbing the dark winding staircase and when passing labourers who were pressed up against the walls to give way. The phrase was also written calligraphically as a decorative insignia on a variety of surfaces including motor vehicles, as well as on the exterior of buildings to protect the structures and their occupants against harmful *djinn* and the evil eye, and was most often found inscribed above door lintels.

Above the black base, the transforming geometries of the future, multi-faceted, forty-meter-high brick tower would radiate from, and be organised by, the centripetal quality of the axial metal rod. All geometries and measurements of the interior and exterior wall surfaces would be defined by a series of calculated points marked by knots and nails along a rope pivoting radially about the metal rod. In the context of Sana'ani minaret design, geometric possibilities manifest in multiples of four which are expressed in the bevelled geometries of the brick tower. Subsequently, the geometries of the tower serve to bridge the four sides of the earth-bound base and the dome which caps the structure and echoes the celestial spheres above.[44]

New steps were added daily during the construction process, circumbulating counter-clockwise around the centre, and keeping pace with the growing height of the wall and central column. With every successive course of stonework, and then later brick, the minaret was extended vertically, limited only by the laws of gravity, yet constrained horizontally by the discipline imposed by the axial geometries prescribed by the builder. Once completed, the vectors of its religiously imbued meaning would transgress its physicality and extend infinitely towards the heavens, but would be nevertheless constrained horizontally on the earthly plain by the audible range of the call to prayer and its visibility as a marker in the urban landscape.

II. Labourers and Labouring

The Bayt al-Maswari and Labour Resources

'Ali Said al-Maswari, father of the present head of the Bayt al-Maswari builders, began his career as a labourer working for several of the great Masters in the 1930s. His successful apprenticeship and skilled craftsmanship led to his recognition as an *usta* by the Masters under whom he trained, and eventually to his own independent commissions. He trained his four sons born of his first wife in the trade, two of whom have attained considerable status amongst the builders of Sana'a. In 1980, while working on the construction of a mosque in one of the newer quarters of the city, 'Ali Said and his sons were offered the commission to erect the minaret as well. This was to be their first. They relied heavily upon the stylistic and structural precedent of the old city minarets for inspiration, and particularly that of the celebrated Masjid Musa, built in 1747–8 AD and considered by many to be quintessentially 'Sana'ani' in proportion and style.[45] Since 1980 the Bayt al-Maswari have been responsible for the construction of some twenty-five minarets in and around Sana'a, successfully cornering the market for these prestigious commissions, and having no other significant competitors to date in this specialised niche of their trade.

Since 'Ali Said's death in 1986, the builders have been headed by Muhammad, his eldest son, and Muhammad's brother Ahmad who is second in command. Each has one of his own sons actively involved in the family business, and one of these has a son working alongside him, representing the fourth generation of al-Maswaris now in the building trade. Muhammad, Ahmad, and Muhammad's son Majid are all *asatiyyah* and the hierarchy of authority on site is ranked accordingly. A fourth *usta* working with them since the early 1980s, 'Abdullah 'Ali al-Samawi, is not from Sana'a but apprenticed under Muhammad and Ahmad after coming to the capital city from a village to the south. Their decision to train an apprentice from outside the family was reportedly due to a lack of interest in the trade shown by the younger family members. Majid's eldest son, Osam, for instance, shared the strikingly chiselled features of his father and grandfather but harboured no desire to be an *usta*. Instead, he pursued his secondary studies diligently with eyes set on medical school. Majid actively encouraged his son's academic interests, conceding that building takes a serious toll on one's body and one's youthful looks. He hoped that all his sons would pursue occupations less physically straining than his own.

At the time of my study, the chief apprentice to the *usta* was someone unrelated to the family and from the same village as 'Abdullah 'Ali. During my stay there, one of Ahmad's adolescent sons demonstrated a serious interest in the family business for the first time and was placed under the guidance of his father and uncle as a second apprentice. Consequently, he was occasionally granted a favoured status over the chief apprentice despite the latter's many years of experience and seniority. Such decisions were evidently motivated by the older generation's desire to maintain their own bloodline in the trade, and this even despite the boy's apparently poor disposition towards the training process and his problematic relations with his fellow workers.

The short supply of a qualified and willing manual labour force was not restricted to the roster of the Bayt al-Maswari, but was endemic to the native population of Sanaʿa and the other large cities in the country. The younger generations of urban Yemenis favoured jobs in the business and trading sector as opposed to manual labouring, and the considerable numbers of young men and women attending university and trade schools in the cities were acquiring qualifications for professional, governmental, and service sector employment in the country's diversifying economy. In these circumstances the traditional building families in Sanaʿa were forced to seek their labour force from the rural populations in the surrounding highland region. In many instances Master Builders heading projects in the capital were also from smaller towns and villages and resided in Sanaʿa only for the duration of their work contracts.

At the intersection of two main traffic arteries near the old university campus, throngs of anxious young men from the countryside lined the sidewalks, trade tools in hand – shovels, hammers, paint rollers – waiting to be picked-up in the back of a prospective employer's truck. Many of these day-labourers had returned from Saudi and the Gulf countries following their expulsion during the Gulf War with valuable trade skills in modern building construction. However, few, if any, of these men found employment with the traditional builders in Sanaʿa, and instead satisfied the labour demands at the sites of new office towers, apartment blocks, and suburban villas being erected on the fringes of the urban sprawl. The labour teams commanded by a traditional craftsman, if not from his own immediate family, were typically from the craftsman's own village in the central and southern highlands, or, at the very least, came from his social web of close contacts.[46] Therefore, not surprisingly, bonds amongst the traditional labourers tended to run comparatively deeper than those between day-labourers on the modern construction sites, as did the sense of obligation and reciprocity. As an employer, the traditional craftsman offered lengthier, more secure contracts, and employees tended to be more conscientious of their work ethic and of the identities they projected to one another.

The corps of labourers working for the al-Maswaris were mainly from the same village located south of Sanaʿa, and about one hour's drive west of Dhamar (see plates 23–28). On average the team consisted of about ten labourers, not including the two apprentices and the higher ranking builders on site. As one young labourer handed me two buckets full of

PLATE **23**

PLATES **23–28**
Portraits of various labourers working
for the Bayt al-Maswari.

PLATE **24**

PLATE 25

PLATE 26

PLATE 27

PLATE 28

mortar, I asked him why all of his fellow labourers were from the region of Dhamar and not from Sana'a. Another of our mates within earshot shouted down playfully from the spiralling staircase above that "It should be obvious by now ... Dhamaris are clearly the best!". I quickly learned that there was a rivalry of sorts between Sana'anis and Dhamaris. Some Sana'anis condescendingly referred to Dhamaris as *"fi'raan al-ard"* ('rats of the earth'), complaining that their vast numbers have infested the capital. The strongly 'labouring-class' image of the Dhamari population lent itself to slighting insults such as *"Hal ant adami wila Dhamari?"* ("Are you human or a Dhamari!?"). But likewise, Dhamaris maintained their own cache of slurs about their Sana'ani rivals. Tim Mackintosh-Smith recounts a humorous tale of this long-standing rivalry:

> The men of Dhamar are the canniest in Yemen and, proverbially, a Dhamari is worth two Sana'anis. There is a story behind the saying: some years ago two Sana'anis and a Dhamari were travelling together. In those days, cotton sleeping bags were used to keep out the cold and fleas. During an overnight stop the Sana'anis decided to play a trick on their companion and, as he slept, they burned holes in his bag with coals from the waterpipe. The Dhamari did not appear to stir but realised what was going on, and when the Sana'anis were asleep he slipped out of bed and cut off their donkey's lips with his *jambiyah*. Next morning, the Sana'anis roused the Dhamari, shouting, 'Look! Look! The stars have fallen and burned holes in your sleeping bag!' 'I know' he replied sleepily. Then he pointed to their donkey: 'Even the donkey's still laughing about it'.[47]

A Thumbnail Sketch of a Labourer's Life in Sana'a

Many of the labourers, *shuqqat al-'imarat*, working for the Bayt al-Maswari had been employed by them for several years, often residing and working in Sana'a for a few months at a time, then returning with earnings to their families in the village for a week or longer. While in Sana'a all of the labourers boarded together. This was a common living arrangement for the majority of migrant workers, especially for those without male relations residing in the city.[48] Their shared rented quarters consisted of a small courtyard attached to the west side of a mud-brick house around which there were three small sleeping rooms and a toilet out-house. Conditions were cramped and rudimentary with three and more men lodging together in rooms measuring no more than 2.5 by 1.8 meters, and sleeping on the floor on wafer thin mats which were covered by tattered, coarse wool blankets. This basic comfort came at a suitably low price of three hundred Yemeni riyals (less than $2.50 American) per person, per month, which

PLATE 29 The author with his fellow labourers.

even the builders considered to be a bargain. Only 'Abdullah 'Ali, the fourth *usta*, had a metal spring cot on which to prop his mattress, and one of the labourers slept on the floor beneath him in order to maximise the use of the available space.

Glossy advertisements for sports cars and Rothmans cigarettes torn from magazines decorated the walls and competed for space with small posters containing Qur'anic verses and a photo of Mecca. Images of Manhattan were misconstrued by some of them as an American sister-city of Sana'a with a glittering concentration of 'tower houses'. The few articles of clothing were hung about on wall hooks, and some of the photos that I had taken of the young men, posing proudly on the work site, were also pinned up. Electricity was supplied to the sleeping rooms only, consisting of single bare bulbs hung from exposed wires. There was no sink or bathing facilities in the out-house, but fresh water was supplied sporadically from a single faucet protruding from the wall next to the latrine. Proper baths were taken weekly at the local *hammam* on Friday mornings, the only day of rest for the builders. There was no space provided in the courtyard for storing or preparing food, which, incidentally, was not something that many Yemeni men would willingly tend to even if there were proper facilities. All meals were taken in small and relatively cheap restaurants in the vicinity of the house or nearby the work site.

The six day working week for the builders was from Saturday to Thursday, interspersed with the occasional national holiday or unexpected day-off due to high winds, heavy rains, or extreme cold. Early morning temperatures below five degrees centigrade are not uncommon during the winter months in the highlands. The daily shift normally began at seven thirty in the morning and often ended after five o'clock, before the *maghreb* prayer, with a short break in the morning for breakfast and a longer one at lunchtime. Although the al-Maswaris did not build during *Ramadan* while I was in Yemen, a common practice amongst builders was to reverse the daily cycle and work through the night until *fajr*, stringing the site with artificial lighting in order to facilitate their tasks. Arguably this practice defeats the true spirit of the fast and is not conducive to either safety or high productivity, but the builders argued that it was not feasible to toil through the daytime without food, water, and *qat* considering the exhausting nature of their occupation.

Conditions on the building site of a minaret were dangerous by almost any standards. There was always an element of risk whether one was teetering on the top of the walls laying bricks fifty meters in the air, passing materials along the steep and dark spiral stair case, or mixing mortar on the ground outside the base of the tower and vulnerable to the paths of free-falling objects that could slip from the hands of those working high above. By mid morning, the dust hung heavily in the shaft, with little means of escape, coating one's throat and numbing one's sense of taste; transforming beams of sunlight from the narrow slit windows into crystalline rays. Debris littered the staircase, and the accumulations occasionally metamor-phosed stair treads into slippery, spiralling ramps. Periodically, several labourers were sent to the site on a Friday morning in order to clean up before the start of a new work week. In fact, however, the presence of debris, the uneven rise of the stair treads, and the empty buckets that were continually sent tumbling down the wind of the stairs to be refilled at the bottom, actually served to heighten one's sense of caution and, as one of my work-mates suggested, might have even reduced the number of potential careless accidents. Admittedly far fewer accidents arose than I had initially anticipated. Many of the labourers chose not to sport footwear while working, storing their Chinese-made rubber flip-flops (*ship-ship*), along with their jackets, on the sill of the slit window nearest their work station. Because of the darkness and the variable conditions of each stair, they tended to place faith in what they could feel beneath their feet, rather than in what they could see, to guide them safely up and down the stairs.

Like all construction sites in Yemen, minaret projects were not off-limits to the general public, and because of the thrilling climb and the rewarding views over the rooftops of Sana'a, they attracted a stream of curious visitors. Elderly men from the neighbourhood made their way slowly to the top, feeling their way along the winding ascent with a cane, and taking a restorative breather at the summit while paying a social visit to the well-known al-Maswari brothers (see plate 30). The only females that visited,

PLATE 30
Two of the neighbourhood men who regularly visited the 'Addil minaret during its construction.

aside from young girls, were a few elderly women, no longer ashamed of squeezing past the strange and perspiring men toiling along the staircase. Upon reaching a critical height, the minaret became the focal point of the local children's urban playground. Every so often half-hearted attempts were made to discipline their rambunctious behaviour, but the inherent dangers of playing on the construction site were never explained to the children in a rational 'cause and effect' manner. The builders adopted the prevalent Yemeni position that 'explaining' to children was a pointless endeavour as they were thought to have no *'aql* (reason). One morning, after three adolescent boys entered the tower and stole a full day's wage from one of the labourer's coat pocket, a sign was posted outside forbidding access to all visitors. By that afternoon, the incident had been dismissed and children were once again squirming past us as we struggled to haul our heavy loads of materials.

Builders' wages, *ujrat al-'ammar*, like the wages of nearly all craftsmen, were once stipulated in the Statute of Sana'a, or *Qanun Sana'a*, which was

91

based on a text of 1748, and was updated by order of 'Abdullah b. al-Mutawakkil (1815/16 to 1835/36 AD).[49] At the time of my study, however, wages could be expected to vary from site to site, as did some of the benefits allotted by employers. The daily wage in 1997 for a labourer at the minaret was five hundred riyals, about four American dollars, but some *asatiyyah* in Sana'a were paying their workers up to six hundred and sometimes provided breakfast as a bonus. Serjeant & Isma'il al-Akwa' report that the daily wage of a builder-in-chief, or *al-usta al-kabir*, in 1972 was fifty riyals[50] – a fraction of the numerical value of an usta's salary in 1997, but nevertheless commanding more in terms of actual buying power and value on the international money markets. Salaries were paid weekly every Thursday afternoon, and predictably spent on necessities, such as food, and supporting habits: notably *qat*, chewing tobacco, and cigarettes. After the rent was paid, the meagre remains were occasionally put toward the purchase of a new article of clothing, a ticket to see an (usually Indian) action-packed karate film at the cinema, or sent back to the worker's family in the village.

The daily schedule was tightly invested with oral activities: smoking cigarettes and chewing tobacco during the morning shift served as a prelude to lunch and, ultimately, chewing *qat* in the afternoon. Lunch inevitably entailed *saltah*, and very often at 'Ali's Restaurant:

> 'Ali himself stands in a cloud of smoke on a platform high above the ground, ladling beef broth, eggs, rice and peppers into a row of stone bowls ... The bowl of *saltah*, as they call the mixture, is brought to you red hot, carried with a pair of pliers and topped with a seething yellowish-green dollop of *hulbah*.[51]

Fare was cheap, normally not more than fifty or sixty riyals per head (forty to fifty cents US), and an additional twenty-five riyals for a bottle of Canada Dry. Seating amongst the builders, and the extension of invitations to paddle into each other's *saltah* with bread, tended to loosely reflect the hierarchy (and gerontocracy) at the work site: the *asatiyyah* huddled around their own cauldron, often joined by the head apprentice and several of the older labourers, while the younger workers gather in one or two separate groups. Occassionally lunch was also prepared at the work site (see plate 31).

After eating, an event that was always accomplished with great haste, the fiery saltah was digested with a glass of sugar-saturated tea, flavoured with a sprig of mint, and followed immediately by a venture into the *suq* to buy *qat* for the chew during the afternoon shift. *Qat*, a tender leaf with a slightly bitter taste and chewed for its mildly stimulating effect, was purchased communally, and every builder contributed his share of the expense which rarely exceeded one-hundred-and-fifty to two hundred riyals per person – approximately one-third of a labourer's daily wage. *Qat*

PLATE 31 Preparing lunch at the work site.

consumption is mentioned by Jamal al-Din ʿAli in his mid-eighteenth-century text, *City of Divine and Earthly Joy*,[52] and the regulation of its sale is stipulated in the *Qanun Sanaʿa*.[53] Shelagh Weir writes that "before the civil war of 1962–70 only a relatively affluent, mainly urban, minority were regular *qat* consumers, but during the 1970s *qat* consumption spread throughout the population and for the first time became a majority practice."[54] At the time of my study, it was popularly believed throughout Yemen that *qat* relieves fatigue and increases one's concentration and productivity. Other wondrous properties, including the ability to relieve minor ailments such as hiccups and the common cold, were attributed to this unassuming leaf which might easily be mistaken by the uninitiated as nothing more than hedge clippings. Headaches, for instance, were dissipated by simply pasting tender leaves on either temple while continuing to chew. Chewing at the building site began when all had returned from lunch and were seated together in a shady spot below the minaret, waiting in anticipation for work to begin again at two o'clock. The *qat*, chewed and stored like a ball in the space between the back teeth and the cheek, was normally enjoyed well into the early evening and often induced a diminished appetite for supper as well as for sleep. I, on the other hand, consistently suffered the abnormal craving for a half barbecued chicken after chewing, and the issue of sleep remained unpredictable from one time to the next.

Aptitude, Motivation, and Aspiration

Majid's son, the fourth-generation representative of the al-Maswaris at the site, joined the work team only in the afternoons after secondary school classes were finished for the day. During my study, he was the only labourer attending school and working on a part-time basis, and was perhaps able to do so since he was also the only labourer native to Sanaʿa, and both his school and his family's business were accessible to him. It was clear that his long-term career interests laid outside the building trade. Both his father and grandfather encouraged his formal schooling and hoped that he, as well as his siblings, would continue their studies to a university level, and they did not wish the back-breaking hardships of their profession upon the younger generation of their family. Simultaneously, and contradictorily, the older generation was also hoping that some of their male progeny would devote themselves to preserving the al-Maswari legacy in Sanaʿa's ancient and distinguished building trade. The two eldest *asatiyyah*, Muhammad and Ahmad, as well as their two builder-brothers, never attended formal schooling during their childhood when Yemen was still under the rule of the Imamate, and they only ever received a few years of basic Qurʿanic instruction. The two younger *asatiyyah*, Majid and ʿAbdullah ʿAli, had never completed their primary school education.

As for the Dhamari labourers, all those in their mid-twenties or younger had received some formal schooling, but none of them had completed their secondary education, and many had abandoned their studies long before ever reaching that level. As a result of the different levels of formal education obtained amongst the builders there were evidently varying degrees of literacy on the site, and more interestingly, the co-existence of different types of analytic and other cognitive skills being employed to comprehend and perform tasks. The analytic skills employed tended to reflect the type and degree of exposure to maths, science, and history subjects, whether in school or during the course of the apprenticeship. As Bloch importantly points out, there is a marked difference between propositional-type, linguistic-based knowledge which is closely associated with schooling, and non-linguistic forms of knowledge which are "formed through the experience of, and practice in, the external world".[55] Those builders who had experienced a greater immersion in the formal schooling process were more apt to supply me with verbal accounts of their trade knowledge than those who had not, but it is important to recognise that their rendition of this non-linguistic knowledge into language entailed a transformation in the character of the knowledge (*ʿilm*).[56] This will be considered in further detail in chapter three. It was also observed that there was an apparent inverse correlation between the individual's level of formal education received and the degree of motivation they exhibited toward mastering trade-related skills and adopting the dispositions (or *maʿarifa*) of

the Masters, and more especially in their aspirations to remain professionally within the trade. It could be generally deduced that those with a high
motivation to faithfully subscribe to the training programme, and with
aspirations for long term involvement, were amongst those who had left
school at an earlier age and had fewer employment opportunities available
to them, but, importantly, were also more ammenable to this type of largely
non-verbal training. For all the young men, long-term planning and goal
setting remained vague, but it was generally wished by the more ambitious
builders that they would be taken on as apprentices (or head labourers)
after five or more years of labouring.

Equally there were many from amongst the work team who were clearly
labouring for financial reasons and not because they harboured any
inclination toward mastering the building trade. One labourer was stashing
away his savings in order to rent a market stall with his brother, and to one
day become a *qat* trader. He boasted: "I will earn 1500 riyals per day, chew
my own (*qat*), and drink Canada (Dry)". Another of the younger boys was
forced to abandon his secondary studies in order to support his widowed
mother and younger siblings. His evident disdain for manual labour and
clear lack of motivation tainted his relations with the Master Builders, and
led to his increasing alienation from his fellow workers. Motivation and
aspiration, when combined with aptitude for performance, were the major
factors that determined the course and success of the individual's
apprenticeship and thus shaping the path leading to his professional
enlightenment. A member of a Sana'ani family of carpenters who have
trained apprentices in the past, told me that "A good apprentice must be
attentive to what the *usta* is doing; must be intelligent enough to understand
the craft intrinsically and replicate the task and respond to given situations;
and must show initiative and truly desire to pursue the craft as his
vocation".[57]

These three factors – aptitude for performance, motivation and
aspiration – are not unlike the three dimensions of the Islamic path of
enlightenment, *islam*, *iman*, and *ihsan*. Both sets prescribe a route for
individual fulfilment, the first leading to a mastery over trade skills and the
accompanying expert discourse, and the latter to personal and spiritual
realisation. Each component in either set represents a level and each level
must be mastered and thoroughly incorporated into one's performance and
thought processes in order to travel 'upward' along the path, and transcend
the lower to the higher dimensions of the Faith. In the context of minaret
building, the metaphorical understanding of transcending occupational
levels in terms of 'moving upwards' is particularly appropriate: the Master
Builders work at the top of the vertical structure, the apprentice(s) directly
below preparing building materials and tools for them, and the labourers,
at the lowest rung, are distributed along the winding staircase, moving
supplies to the top in assembly line fashion. The work site, as *opus*

operatum, is in Bourdieu's sense that 'book' from which the builder learns his vision of the world: a book "read with the body, in and through the movements and displacements which make the space within which they are enacted as much as they are made by it".[58] Learning is cumulative and each step, or *object of knowledge* mastered, informs one's approach to, and understanding of, the next in a structured fashion. Titus Burckhardt writes similarly on the teaching methods of an instructor at the Qur'anic University of al-Qarawiyyin:

> Mulay 'Ali was against any overdue haste. He did not permit an author to be referred to hastily or without the student wishing him God's grace, nor did he allow anyone to anticipate the proper logical development of a thought. Each brick in the edifice of a doctrine had first to be sharply cut and polished before the next one was added to it.[59]

As Bourdieu points out, the co-ordination of practices, and the practices of co-ordination, as well as the young builder's capacity to carry out corrections and adjustments to these, presupposes "their mastery of a common code" which may be defined as the *habitus* of the group. The dialectical relation between the body of the builder and the strictly hierarchised space of the building site lends itself perfectly to a structural apprenticeship which results in "the em-bodying of the structures of the world". In turn, it is these structures which facilitate the mastery of co-ordinating practices (and practicing co-ordination), and render unnecessary the memorisation of each and every procedure in the cumulative process of becoming an accomplished builder.[60] Likewise, Bloch postulates that in becoming an expert one does not simply "remember many instances", but rather constructs "a dedicated cognitive mechanism for dealing with instances of a particular kind" which can relate to a specific domain of activity quickly and efficiently.[61] It is through the builder's regular immersion in the practice of closely related activities that he develops such a purpose-dedicated domain,[62] and effectively comes to embody the *habitus* of the traditional Sana'ani builder.

Discipline and Religious Training: Islam and *Islam*

Islam, the first and crucial step in becoming a Muslim, is about submission and it is this dimension which lent its name to the religion. The main characteristic of this dimension of Islam is performative and is premised upon performing the rites, or pillars, of Islam, namely voicing one's belief in the unity of God (*shahadah*), daily prayers (*salat*), fasting (*sawm*), the pilgrimage (*hajj*), and paying the religious tax (*zakat*). Seyyed Hossein Nasr draws a parallel between Islamic spirituality and Islamic art, observing that "the Islamic rites mould the mind and soul of all Muslims including the artist and artisan".[63] It is therefore important to have some understanding of how these rites are imparted to children, both in the home and within the school curriculum in Yemen, in order to gain a better insight into the nature of the particular teaching-learning process, and to determine what repercussions it has on the traditional apprenticeship system.

Religious training begins in the home, commencing from the moment the *shahadah* is whispered into the ear of the new born child. This marks the initiation of a process of "internalising meanings".[64] Vom Bruck notes that in Yemeni orthodox families, "parents ensure their infants know the name of God before they teach them kinship terms".[65] It is not until the age of six or seven, however, that a conscious effort is made by the parents to inculcate the performative aspects of the religious rites in the child. Most Sanaʿanis agreed that there is little point in imparting a structured education and a sense of self control at an earlier age as children are thought to have no *ʿaql*, reason, before seven years of age.[66] Parents who displayed exasperation in trying to reason with their young children were thought to have lost their own sense of *ʿaql*. Adults in Yemen were expected to display a high degree of tolerance in the face of unruly behaviour from little ones, whether their own children or others.

One evening while walking along the top of the banks of the *Saʾila* (the dry river bed which divides the old city from the Bustan al-Sultan quarter immediately to the west), I watched with great amusement as a guard protecting the *Saʾila* renewal project struggled in vain to keep droves of little children from running barefoot through a patch of fresh concrete. He menaced the children rather playfully, wielding a stick and shouting ferociously. One group of children would scatter, screaming as they scurried to safety, and another group would close in behind, gleefully leaping into the concrete. The guard apparently enjoyed the naughty game as much as the children, typically with no intention of truly disciplining the toddlers. Older children who were expected to know better were fully aware that they would be in serious trouble if they partook in the shenanigans of their younger siblings. This widespread notion that 'reason' is absent in young children is reflected in Vygotsky's theories of a child's conceptual development, in which he claims that although by school age children have developed an array of conceptual representations which approximate adult forms in their outward appearance, these are still not subject to conscious awareness, or actually part of an organised system. In order that the processes of conceptualisation come under voluntary control, the child must be subject to a training involving highly reflexive activities.[67]

The first of these activities to be taught in the course of the Muslim child's religious training is prayer, and it is the parents of families who practice their Faith who are the principal agents of indoctrinating rites, as well as being the teachers of language, social customs and etiquette. Generally, it is the mother who figures most prominently in socialising the latter three, whereas the father plays the more active role in teaching, especially his sons, about their religious rites and duties. At this early stage, learning to pray initially involves the mastering of the bendings and bowings involved in prayer prostrations, and later, the correct performance of ablutions which are necessary for validating prayer. Children who are already accustomed to watching their parents and older siblings perform prayers, are now requested to make concerted efforts to watch and imitate if they have not already taken the initiative to do so themselves. The *Sunna* and the *Hadith* instruct parents on how to teach their children, and these Islamic sources also recommend that a father coerce his children to correctly replicate the rites and to be dutiful in their daily prayers, using strict disciplinary measures if necessary. This period of coercion may endure for up to five years after which time it is expected that the child will have developed *'aql*, or reason, and should be responsible for upholding their own sense of religious duty.

The ability to pray was thought by neighbours in my *harat* to be directly associated with the development of reason,[68] and some young boys began to assert their independence by going to the mosque with an age mate, unaccompanied by their fathers. Gatherings of young boys occasionally mimicked the call to prayer in game, howling the words in unison with the *muezzin*, while proudly displaying their knowledge of the recitation to friends. The ambition of becoming so learned in Islam so as to one day become an *imam* was common amongst young boys, although very few would actually ever pursue this course in life. To the minds of children, an *imam* is rightly conceived as a patriarchal figure commanding great respect in the community. At a level presumably more complex than the child's understanding of power, religious scholars have maintained a prevalent position within Yemeni society as key players in the Islamic discursive tradition which moulds the identity of the community and its present practices in relation to conceptions of an Islamic past and future.[69] They also continue to influence the nation's political climate, and, to a certain extent, control the production and reproduction of social values and popular customs, particularly amongst the more conservative sectors of the population.

These physical performances are accompanied by other disciplinary measures, such as waking early and regulating the course of other daily activities in accordance with the calls to prayer. The combination of performance and discipline forms the foundations of Islamic socialisation, acting upon the trainees' conceptual understanding of themselves in strong relation to their environment: socially and spatially in terms of how, with whom, and where they enact their rites; and temporally in terms of structuring the schedule of their daily lives. The prayers effectively spatialize time by regularly punctuating the day and night, and thereby spatialize the course of daily activities in rhythm to the religion. Therefore, prayer in Islam may be understood to have an inherently spatializing structure, not unlike the carefully regulated practices of other religious institutions.[70] Regulated spatialization of the overall environment gives those who subscribe to it a strong sense of 'place', and their thoughts and actions pertaining to all endeavours may be oriented within this conceptualised framework.

Fasting during Ramadan, along with the proclamation of Faith and the daily prayers, are the rites taught in the early stages of a child's Islamic socialisation. Alms giving, pilgrimage, and possibly the concept of *jihad* are addressed much later when the individual has developed a deeper understanding of their religion accompanied by a sense of faith. Conversely, it should be noted that children may be given money by their parents to give to a beggar when passing on the street, instilling an early sense of their *zakat* duties. Likewise, the teaching of right from wrong, and encouraging children to question the moral implications of their own thoughts and actions establishes the groundwork for the individual's dealing with the more complex issues of *jihad*, or holy struggle. During Ramadan, fasting is not normally enforced upon a child in the same manner that prayer is encouraged, since children are not thought to be physically capable. Fasting is normally begun once children demonstrate their own initiative to participate, usually when they are more than seven or eight years old, and stemming from a desire to imitate their mentors. In some Sana'ani families, children were told by their mother to keep their eyes closed while eating at the meal table else "Ramadan" will see them. Ramadan was personified in order to make the idea of this religious holiday tangible to the child's imagination and thereby introduced a playful element to the rite. After the meal, children were told to open their mouths wide and to hold out their tongues so that their mother could pass the dull end of a sewing needle along it, telling her children that their tongues were now sewn down so that they could not snack between meals. In fact if children wanted to eat between meals, parents rarely prevented them from doing so. By playing this game, children sensed that they were participating and parents were satisfied that their children had been properly nourished.

Children are expected to begin fasting for real around ten years of age, but progression begins in an incremental fashion by skipping breakfast, then eventually cutting-out lunch as well. Immersion into full fasting is paced according the child's own motivation, and this is tempered not only by the social environment of their home, but also by their peer group and the influence this has on their desire to emulate the elders. Eventually fasting will be extended for full-day periods, but initially for only a few days per week while the child is still in the process of building physical and spiritual strength, and disciplining their temperament. By the time they have

99

reached fifteen years of age, the individual is considered to be sufficiently prepared, and to have ascertained sufficient reason, *'aql*, to participate fully in Ramadan.[71] This age may be earlier for females depending on the family's ambitions for marrying their daughter at a young age. Fifteen years old corresponds roughly with the point in the formal schooling curriculum when students are taught the deeper significance of the fast, as well as their other rites and duties as Muslims. Nasr writes that Ibn Sina (980–1037 AD), whom he considers to be the foremost among the Muslim philosopher-scientists, believed that "at the age of six, (the child) may be given tuition by a master... who will teach him step by step and in order... Meanwhile, grammar should be taught to the student, followed at the age of fourteen and onwards by mathematics and then philosophy".[72] It is interesting to note that the age which Ibn Sina recommends for the study of philosophy in this passage corresponds with the age set by the scholastic curriculum in Yemen for studying the deeper spiritual significance of Islam. It is expected that an Islamic conceptualisation will eventually supplant the merely mechanical understanding, and come to inform the performative aspect of the rites. Nasr remarks on the subject of fasting that:

> The outward observation of its rules, while necessary, is one thing and the full realisation of its meaning is another... (being) the realisation of the ultimate independence of man's being from the external world and his dependence upon the spiritual reality which resides within him.[73]

Akbar Ahmad, author of *Discovering Islam*, writes that beyond the Faith's apparent and seductive simplicity – one God, one book, one Prophet – and clearly defined rituals, there are far greater depths to be explored by those so-inclined.

> As a boy I thought the 5 daily prayers were meant to instil discipline – the regular washing and waking at early hours in preparations for prayers and the bending and bowing during them. Later, in manhood, I gradually perceived the deeper significance of the prayers. They were a constant reminder of the transient, passing nature of the world. And, related to this, a constant declaration of the permanence of Allah. Muslim prayers can create sublimity around the believer, peace within.[74]

Religious Education and the School System

During the period of the twentieth-century Imamate rule, education outside the capital was, for the most part, restricted to the Qur'anic schools, *madrasah* or *katatib*, staffed by instructors of often questionable levels of literacy. Historically, these institutions found little sponsorship amongst the predominantly *Zaydi* population of Yemen's highlands in stark contrast to the *Shafi'i* areas in the west and south of the country. Rather, "Zaydite scholars taught voluntarily in an honorary capacity", usually at mosques or other sacred places under tribal protection.[75] In Sana'a, however, Ottoman-introduced reforms to education maintained and supplemented by Imam Yahya led to popular talk of the 'new knowledge' (*al-'ilm al-jadid*), describing science, geography, and *belles lettres*. Imam Yahya nevertheless disapproved of this new knowledge which he associated with "the culture embodied by European colonial powers and [he] feared for the moral contamination of his people and their alienation from Islam".[76] At the time of the Revolution in 1962 the Egyptian Government supplemented the approximately one hundred trained teachers in the Yemen Arab Republic with an additional forty-six teachers and five educational experts,[77] and the secularisation of the school system resulted in a reduction in religious instruction to one hour per week.[78] The number of Egyptian teachers in the country grew to sixteen thousand in the mid 1980s, but have dropped dramatically since largely due to the government's incapacity to meet the necessary expenditures, and their subsequent replacement by other foreign nationals with lower wage demands, most notably Iraqis and Sudanese. Not surprisingly, curriculum and syllabus of the national system have been strongly infiltrated by foreign influences, not unlike that of Turkey's whose new educational philosophy stressing 'rational learning' was imported by Mustafa Kemal from France.[79] A series of measures have been taken since the 1970s to tailor these outside interventions to Yemeni needs, and there has been a concerted effort in recent years to bolster the proportion of Yemenis employed in the educational sector.[80]

The national educational curriculum is predominantly secular, but includes compulsory religious instruction from the start of basic school through secondary school, thus more or less absorbing the role of the *katatib*, but not altogether dissolving these institutions. At the time of my study, some families in the city centres continued to send their children for supplementary religious instruction outside regular school hours, usually between *salat al-'asr* and *salat al-maghreb*, whereas other parents protested that their children were being overwhelmed by the heavy dose of religious instruction in the national curriculum. Religious instruction was also co-ordinated with higher studies. All university students, regardless of their programme, had to take an additional course on Islam during their undergraduate studies, which for the majority of students would constitute the final phase of their formal religious tutoring. This final course reviewed the material studied in the previous twelve years of schooling and aimed to challenge the students to deepen both their practical and philosophical understandings (*'ilm*) of Islam. Assignments in essay writing on topical issues forced the more mature students to draw upon the spectrum of religious subjects learned – for example Qur'an, Hadith, jurisprudence, etcetera – and to thread them together in the form of cohesive arguments.

However, as mentioned earlier, none of the builders with whom I worked had even completed secondary studies, nor had any attended Qur'anic school beyond a very elementary level. Basic school, lasting from grades one to nine, was the highest degree of education obtained amongst the rural labourers. In fact, despite a government policy of compulsory education for those between the ages of six and fifteen years supported by a law promoting equal education rights for both sexes, and no tuition payments for basic education, many of my colleagues were forced to abandon the system at much earlier ages. Reasons tended to combine both economics and social considerations. Adolescent boys, often by necessity, sought employment in order to support themselves and to contribute to their family's income. They were actively discouraged by parents and peers from pursuing studies as there were no perceived gains and prolonged education was viewed as an impediment to early marriage. This scenario was typical in Yemen: according to figures compiled in 1991–92, only 57.5% of those between the ages of six and fifteen were actually enrolled in basic school in that academic year, and reasons for non-attendance closely resemble those which I have suggested above.[81]

Therefore, for the purpose of illustrating a correlation between religious training and trade apprenticeship, it is useful to limit considerations of religious instruction to the basic school curriculum, in addition to that imparted by the home environment. During basic school, religious education focuses on the recitation of the Qur'an, learning to perform prayers and ablutions, and instruction about *Hadith*, *Tawhid*, and proper comportment.[82] Religious instruction is taught daily and divided into three subjects: Qur'an, *Hadith*, and *fikrah*. Emphasis is not placed on teaching a catechism, but rather on training the body in terms of behaviour and performance. The ultimate aim is to render religious practices and observances "a natural and organic part of the human configuration".[83] The first year text book instructs the children on how to make proper ablutions and perform their prayer prostrations with the use of cartoon drawings and accompanying descriptions. The curriculum also addresses issues of moral conduct and what is expected from a good student, and son or daughter, in terms of disciplined behaviour. Qur'anic studies begin in the first year with the opening *sura*, The *Fatihah*, and proceed with the final *sura*, *an-Nas* (#114), working backwards through the shorter *suras* of the Holy Qur'an.

The format remains essentially the same for the duration of basic school, though studies of the individual subjects become more involved, especially with regards to hadith and the life of the Prophet which provide models for individual discipline and comportment. By fourth grade there are separate texts for Islamic instruction and for Qur'anic studies, and the former is replaced in the seventh to the ninth grades by a text on the Hadith and instruction on proper upbringing. Increasing emphasis is placed upon mastering recitation and memorising the Qur'anic verses, and explanation of the theological definition becomes more involved. Vom Bruck explains that, traditionally, "the status of memoriser was higher than that of a person who merely knew how to write".[84] It was explained to children that rote learning would develop the mind and facilitate reasoning ('aql).[85] Memorisation and embodiment of the text are implied by the title, al-Qur'an, derived from the verb qara'a meaning 'to recite'. Originally, Muhammad did not write the book but recited it, and, subsequently, Muslims do not read the book but ideally recite the Qur'an.[86] Qur'anic studies continue in the reverse order of the suras, from the shorter to the longer, so that by the end of the ninth grade students have covered the readings of the entire Sacred Book with the exception of suras two to six. Basic school religious training essentially equips the student with a sufficient background in the practical, performative, aspects of Islam to enable the individual to transcend this surface level of enactment to one of deeper faith, or iman, if they are so inclined. In fact, the secondary school programme serves to merely broaden the scope of the students' knowledge ('ilm) about Islam: the decision to embrace the Faith is recognised to ultimately be a personal choice, one which is inspired by individual motivations, and not possibly subject to coercion by an outside agent. Interestingly, many fathers[87] I spoke with about education felt that the curriculum in the government-run public schools was overburdening their children with an overload of religious studies. The content of these courses was often felt to be beyond the comprehension of students at basic school level. It was popularly believed by these parents that the heavy dose of religious indoctrination in the school system was in response to pressure put on the ruling People's General Congress by their main political opposition, the Yemeni Congregation for Reform (Islah).

Conversely, for those parents seeking a more intense religious immersion for their children in the school system, the Yemeni Congregation for Reform, or Islah, have devised an alternative option to the national secondary schools which they have called ma'had al-'alimiya, or scientific institutes. The weighty religious curriculum of these schools reflects the Islamicist ideology of the party and has provided a platform for its radical figures to preach a militant Islamicist message.[88] The popularity of the ma'had al-'alimiya has steadily increased since the early 1990s in response to several factors including the sharp rise in religious conservatism in Yemen. This rise has been fuelled by the expulsion of over eight hundred thousand migrant workers from Saudi after the eruption of the Iraq-Kuwait crisis in 1990, many returning home with the potentially explosive combination of frustration and a taste for conservative religious ideology.

As Buchman points out, many of the country's young people enduring the resulting economic hardships are drawn to political Islamic movements and ideologies which promise outward action and material benefit.[89] Some Islah-sponsored schools offer free board for their students from the villages, marketing a tempting offer to the economically dispossessed, and in general supplying a better option than the poorly funded and badly administered government-run schools. However, with a clear majority win for the ruling party in the country's 1997 elections signifying a decrease in Islah's popular support, the President has found better leverage to reunite the country's school system under the sole jurisdiction of the government. Pressure will remain from the persistently influential Islah figures, and in particular the paramount *shaykh* of the Hashid tribe, 'Abdullah al-Ahmar, to upgrade the quality of the system and perhaps to strengthen the religious component in the curriculum.

The need for upgrading the quality in the government-run schools has struck a chord with all Yemenis. The system is heavily under-funded and the classrooms are over-crowded. It is not uncommon for inner-city classes to exceed one hundred students at the secondary level, and for classroom furniture to be destroyed and left rusting in piles outside in a corner of the schoolyard. Teachers and administrators are expected to adopt extreme disciplinary measures toward their students, running the schools more along the lines of a correctional institute than a place of education. One Iraqi teacher in a Sana'a secondary school recounted how once, when his normally unruly class of over one hundred boys was completely out of control, he resignedly left the classroom refusing to teach unless they changed their behaviour. A colleague seeing him in the corridor, inquired as to what had happened and decidedly took it upon himself to discipline the class. He marched in and stood at the front where there had once been a chalk board, and verbally abused the entire class, yelling at the top of his lungs "we know you are sons of dogs ... sons of whores ... ", until there was complete silence. Mission accomplished, he left the room, turning the now quiet class over to his colleague who stood waiting in the corridor. The teacher re-entered, calmly pausing before announcing in a disgusted tone that he would *now* refuse to teach because they were "a worthless group with no self-esteem to have put up with such insult", and again left the classroom. He complained that his attempts to reason and treat the boys as adults were completely disregarded by both his colleagues and his students. Some students were bold enough to accuse him of being spineless and having no personality because he did not display aggressive tendencies toward them. Undeniably, aggressive assertiveness constitutes an esteemed quality in Yemeni males, and is also expected to qualify the disciplinary measures taken in any training process.

The Builders and Islam

Due to basic school training, to whatever extent received, and socialisation within the home environment, the moral constitutions of the minaret builders, without exception, were shaped by an Islamic tradition supplying a rich resource for personal expression. This tradition was drawn upon in the everyday spoken use of traditional expressions and religious proverbs. The appropriate phrases were mechanically issued in response to the specific situation and context, and the majority of these invoked peace, blessings or the Name of God: formalised greetings and salutations corresponded to the time of day and the relative status of the speakers. For example: *"massakum Allahu bi-l-khair"* (May God give you a good evening), to which the response is *"massaka Allahu bi-l-khair wa-l-'afiya"* (May God give you a good evening and good health); or bidding farewell, *"ayy khidamat"* (Any service), to which the response is *"Allah yahfadhak"* (May God preserve you); blessings on someone who is sick, *"Allah yushafeek"* (May God restore your health); on someone who has received bad news, *"Allah yustur"* (God protects); blessings on someone who has sneezed, *"yirhamak Allah"* (May God have mercy on you) to which the reply is *"yirhamna wa yirhamakum Allah"* (May God protect us and you); or burped, *"ya hana'"* (good health), to which the reply is *"hannak wa ma hanneeni hannak"* (good health to you, and what gives me good health is your good health); or who has just had a bath or a haircut, *"na'iman"* (Good living), to which the reply is *"Allah yin'am 'alaik"* (May God grant good living upon you); exclamations in response to a noteworthy or astonishing event, *"ma sha'Allah"* (what God wills); to the bereaved on the news of a death, *"dhamman Allahu ajalakum"* (May God guarantee your appointment with death); and so on. Nasr remarks that "the continuous references in the Qur'an to the eschatological realities and the fragility of the world, (and) the constant repetition of Qur'anic phrases ... remould the soul of the Muslim into a mosaic of spiritual attitudes".[90]

Likewise, their physical comportment was also patently fine-tuned by their socialisation as both Yemenis and Muslims within this particular cultural context. Lunchtime provided the best opportunities to witness these standardised daily performances. A by no means comprehensive list consists of the following examples: relieving oneself around the work site involved choosing a location that would afford maximum privacy and discretion, and taking a squatting position when urinating was deemed necessary in order not to splash one's clothing and make them impure (for prayer); washing one's hands was necessary before sitting down to eat; eating and passing food onto others was executed with the right hand only; it was the norm to share food communally from a common plate or bowl and to offer food to any person who joined the long communal eating tables in a restaurant; a burp was intercepted by placing a cupped right hand to the chin, fingers pointing upwards, and blowing any potentially foul-smelling air upwards and away from those with whom one is speaking; males walked hand-in-hand with each other as an outward expression of their friendship and the strong male bonds propagated by their culture; social contact with females, including direct eye contact, was avoided under all normal circumstances, unless she was blood related; etceteras.

In contrast to this general behavioural code premised on discretion, hygiene, and good will toward one's neighbour, a considerable amount of behaviour on the work site countered any ideals of Islamic 'good' comportment. This was somewhat surprising, not by the fact that it revealed inconsistency in character or behavioural patterns, but rather that it occurred, and was tolerated, on a building site directly associated with the sacred precinct of a mosque. Aside from the more predictable occurrences, such as an occasional shouting match, minor scuffle, cursing, burping, smoking, spitting, farting, or a lewd joke, some of the behaviour was clearly more questionable in terms of its social acceptability. Several of the builders frequently leered at, and made suggestive comments about, any woman passing the minaret site. This seemingly universal pastime of construction workers had to be carefully tempered in Yemen where crossing the line may result in serious consequences: men caught looking lustfully at women in public places during Ramadan may be arrested and their heads may be shaved as a penal sign for public consumption.[91] Those who did partake in this activity at the minaret tended to do so in a very exaggerated way, obviously determined to make their masculine and heterosexual intentions more apparent to their fellow workers than to the passing women. From high above, at the top of a minaret, the object of desire was normally only visible as a billowing mass of black fabric passing on the street some forty meters below. The activity was more heavily invested with bravado and with assertions of masculinity in a male dominated environment than with trying to seduce a potential mate. Likewise, homo-erotic horsing-around, invariably initiated by a more senior builder and directed toward the younger labourers, translated clearly into male relations of power and domination in an arena of competing masculinities, serving primarily to reinforce the existing hierarchy.[92]

In short, the religious understanding of the average labourer did not extend beyond a fairly elementary knowledge of the *furud*, or religious duties, which oscillate between the *halal* (that which is permitted) and the *haram* (that which is forbidden). Between these two extremes lie those things that are *manduub* (that which is recommended), *mubah* (that which makes no difference), and *makruuh* (that which is reprehensible). It was largely by these terms which they passed moral judgment on the actions and habits performed by themselves and others. The builders' socialised sense of

Islam was expressed through the particularities of their unquestioning beliefs, their performances, their abstentions, what they said, and what they did not. In relation to this ordered grid of specified parameters, they located their moral selves in terms of what it was for them to be a Muslim, and what it was to be Yemeni, and likewise manipulated their identity, both as they believed it to be and as they wished for it to be projected, within a framework which was perpetually formulated and reformulated by the dynamic tension between particularistic and universalistic concepts of Islam.[93]

Their grasp was of the *particulars*, each practice as a discreet object of knowledge, exercised automatically in response to a formalised set of circumstances. The relations between the circumstances and the actions (whether embodied in physical performance, voiced, or conceptualised) represented the elemental components of their Faith, the *particulars*, the concrete connections that bridge their concepts of the self with concepts of the eternal. More profound understandings of those connections require an ability to transcend the simple mapping of 'A gives rise to B', and forge connections of greater magnitude between formerly disparate ideas through a process of metaphorical abstraction. Drawing the metaphoric connections effectively charters the map of one's Being, thereby expanding the boundaries of conceptual space and forging increasing connectedness between all realms of thought and action, thus moving one closer to *tawhid*, the ideal oneness of God. This necessitates an appropriate training and ample time for meditative reflection. Not surprisingly, none of the young labourers had been equipped to make that leap into the fathomless oceans of their Faith.

Nevertheless, while on the job, two qualities in particular impressed me as being characteristic of my fellow workers, as though engraved upon their dispositions by earlier processes of socialisation and discernible in the general attitudes they exhibited. The first was the general abstention from physical fighting and the control of emotions, in particular those associated with rage, jealousy, fear, and sadness: in other words, a command of their *'aql*, or reason.[94] Losing control of one's temper in public was perceived as self-denigrating and childish. The few skirmishes that I did witness were resolved quickly, and in one case, despite a black eye and a generous dose of bruises, the adversaries were once again companions the following day. The second quality, which I will discuss in the following section, was the seemingly unquestioned acceptance of authority which was manifest in the rigid hierarchy of the building team, and which ultimately controlled the knowledge and power of the trade.

Basic knowledge of the Qur'an, the life of the Prophet, and other religious stories was most apparently expressed in the lyrics of some *hajl*, or labouring songs, which the builders sang in unison as a work team.[95] Many of these *hajl* also celebrated agricultural and harvesting themes which played an integral role in the men's rural backgrounds in Dhamar. The city-dwelling Bayt al-Maswari were not so well-versed in singing *hajl* in comparison with their rural colleagues. This left open the opportunity for the more senior Dhamari workers to jest that they possessed knowledge of at least one important thing which they could teach to their Masters. The song master, very often the *thana'a* (second in command), did not hesitate to correct those who confused lyrics or mispronounced the sometimes obscure words of the songs. In other cases, some of the men rhythmically recited Qur'anic *suras*, in part or whole, which they had memorised during their earlier schooling. Like the *hajl*, the recitation of *sura* served almost entirely to regulate the tempo of their physical labour, and not to induce spiritual meditation.

None of the builders had delusions of being Qur'anic scholars. In fact, only two of the builders prayed regularly, or at least claimed to do so with some credibility. Noting that none of the labourers prayed the *salat al-zuhr* in the mosque adjacent the minaret site during the lunch break, and that there was no interruption in the work schedule for the mid afternoon *al-'asr* prayer, I began questioning my fellow workers about their religious practices. Initially the universal response was that they were all pious Muslims and prayed five time a day as decreed by orthodox doctrine. These early answers were not altogether surprising, especially when considered retrospectively as most of the young men had expended a great deal of energy in finding out what religion *I* subscribed to and why it was *not* Islam. Some made minor attempts at proselytising and explaining how much better off I would be, especially in view of the long term scheme of things, if I would make the conversion to the Faith.[96] Seemingly, at this early stage in our acquaintance, they could not afford to appear to be neglecting their most fundamental duty as Muslims. They were, in effect, constructing a projected identity by carefully negotiating their own positions within the parameters of their Faith, and in accordance with a set of anticipated expectations that they cast on me as the outsider, the Westerner, the non-Muslim, the *nasrani* (Christian).

Increasingly, tea breaks were transformed into stimulating periods of exchange, both myself and the builders curiously prodding each other with questions about our respective lives, backgrounds, and societies. The subject of alcohol frequently cropped up on their agenda of interests, and I found myself occasionally being lectured to in a stern moral tone on the blasphemies of drinking. Having seen enough empty Johnny Walker bottles around town to know better, I suggested to my colleagues that they were possibly better acquainted with alcohol than they admitted. This provoked blushes in manifold shades and some nervous laughter, and, later, hushed divulgence about fellow colleagues or senior builders, but never accounts made in the first person. In theory, drinking alcohol (*khamr*) remained *haram*, but in practice the full gradation between *haram* and *halal* became fully apparent.

Several months later, during a lunch break, a small group of militant Islamic extremists on route to the mosque for *al-'asr* prayer began casting stones at several of the young builders who lied sleeping in the shade at the base of the minaret. The antagonists angrily accused them of neglecting their prayer duties and challenged their status as Muslims. Following this peculiar incident, I was once again prompted to inquire about their religious observances, and at this point in time I had established comfortable relations and friendships with many of them and was more confident about the real truth of the matter. I received very frank, almost disinterested responses from the majority of them, confessing that they "rarely, if ever, prayed", accompanied by a casual shrug of the shoulders. This markedly altered response may have been partially an act of resistance to the rock-throwing extremists and the oppressive wave of religious conservatism which they represented, but it was also more in line with their actual practices. Only the head of the Master Builders, Muhammad al-Maswari, and his *thana'a* claimed to pray five times a day, attesting that missed prayers at *salat al-zuhr* and *salat al-'asr* were compensated by additional prostrations during evening prayers. Some of the other *asatiyyah* also claimed to pray, but admittedly on a less regular basis.

Training and Discipline

In contrast to the generally slack attitudes displayed in connection with religious practices and observances, there was an unequivocally strong sense of discipline with regard to work ethics. This stems from two principal factors which were very much in a dialectical relation with one another: the authority of the Masters and the (untamed) will of the labourers. The Master Builders, and particularly the head *usta*, were strict with their work team and operated in a clearly autocratic fashion. The will of the (new) labourers necessitated harnassing in order to synthesise an efficient team which complied with orders and expectations. This disciplined teamwork was largely the product of, what Bourdieu would recognise as, the 'objective homogenising of the group *habitus*', resulting from the "homogeneity of the conditions of existence [which] enable paractices to be objectively harmonised without any intentional calculation or conscious reference to a norm".[97] The labourer's submission has a strong historical precedence in the apprenticed trades, as well as in the Sufi orders, of the Middle East and North Africa. In discussing slavery during the Medieval period, Amitav Ghosh writes that "equally in some vocations, the lines of demarcation between apprentice, disciple and bondsman were so thin as to be invisible: to be initiated into certain crafts, aspirants had to voluntarily surrender a part of their freedom to their teacher."[98] An additional consideration was that, for all intents and purposes, the individual labourers were replaceable since, in effect, they did not offer their employer any specialised trade skills except their obedience and willingness to perform manual labour. During the year which I worked for the Bayt al-Maswari, only one labourer was led permanently off the site by an angry head *usta* for talking back to his seniors, but this example supplied the others with a sufficient reminder of what the consequences might be for misconduct.

As previously stated, there was a fairly regular circulation among the labourers, many of whom worked in Sana'a for several weeks then headed back to their village for a sojourn with their families and attended to local business, and returned once again to the capital. The same labourers were repeatedly re-employed by the Bayt al-Maswari, therefore producing a carrousel effect of faces, and enabling the men to periodically rejuvenate with the security of knowing there would be work for them when they returned. New labourers were nevertheless taken on from time to time during my period at the site, and not necessarily from the same Dhamari village as the others. Like my own first day on the job, there was no formal introduction of the new member to the rest of the team, and more so, no explanation was offered about the nature of the project, the tasks entailed, or the skills required. New members were merely expected to find their own way into the machinery of the team. Initial orientation was facilitated by the other labourers, and feedback came most often in the form of reprimands usually devoid of corrective instructions. The values of being an efficient labourer were 'made body' by the labourer's immersion in an 'implicit pedagogy' – defined by the site, his engagement in a set of practices, and his (sometimes antagonistic) relations with the other members of the building team – and reinforced through the coercive effects of (normally abusive) injunctions.[99]

In contrast to the construction schedule of most other building projects, that of a minaret is both more linear and more firmly in the control of a single individual. A new house, for instance, involves the scheduling of various sub-trades, sometimes simultaneously, and requires management supervision and the delegation of responsibilities as opposed to a 'controlling' autocrat. The Master Builder for such a project, sometimes in conjunction with the client,[100] oversees the co-ordination between teams of mud plasterers, *guss* plasterers, *gamarriyah* window makers, finishing carpenters, electricians, and plumbers, in addition to his own team of labourers and apprentices who are responsible for erecting the structure. Because of the more complex division of labour, apprentices on a house site normally procure greater autonomy and are assigned tasks demanding greater responsibility than those who were under the tutelage of the al-Maswari brothers.

Muhammad al-Maswari, as head *usta*, did not share the extent of his project plans and schedules with his team members, except possibly with his brother Ahmad, and he parcelled out specific assignments to his son Majid and the fourth *usta*, ʿAbdullah ʿAli as-Samawi. No other trades were involved in the construction of the tower until the structure was completed, at which point the *guss* plasterers arrived to finish the surface on the underside of the staircase and to highlight the exterior decorative patterns in shining white. Throughout the stages of construction, the *thana'a* and lower level apprentice(s) gleaned their respective duties according to what was required of them by the *asatiyyah*. The *thana'a* was generally delegated more determinate tasks and a greater degree of autonomy than his fellow apprentice(s). Below these echelons, the labourers were expected, for the most part, to organise themselves by dividing the tasks and territories which had been defined for them by their seniors, and to evolve an internal sense of hierarchy in which all team members could be located. These parameters, the least codified within the trade hierarchy, were in continual flux, and most apparently when significant changes were introduced to the system such as the commencement of a new phase in the building programme, or the departure of a more senior member of the team or the introduction of a new one.

It was during these periods of reorganisation amongst the labourers themselves, in terms of designating new positions on the site and redefining tasks, that the distribution of power at this level became more apparent (see plates 32–35).[101] Those senior in age, and more often those who were highly ambitious, eagerly perpetuated a system of surveillance upon their fellow workers, ensuring that each member was upholding their share of the labour responsibility. For instance, exact numbers of stairs were counted out along the spiral and designated as the territory for each individual, and scoldings were swiftly dealt to those discovered to be manning less or more than their allotted share. Likewise, individuals felt to be working too slowly or too quickly, or stockpiling materials on their landings and not passing them onwards, thus effectively disrupting the smooth operation of the 'engine's pistons', were compelled by their work mates to re-conform to the working rhythm. The tempo of this rhythm was largely set by the demands for materials from the *usta* above, but occasionally a labourer working at the bottom where the supplies were delivered, or nearer the top where they were stockpiled, would regulate the velocity of the flow on his own initiative.

Competition between the more enterprising labourers became manifest in tests of strength and endurance, as well as in their efforts to obtain the trust and approval of the Masters. Fatigue and injury, whether physical or emotional, had to be carefully concealed in order to avoid a barrage of teasing. Although minor accidents and physical injuries normally elicited laughter, concern for the well-being of others was nevertheless displayed

PLATE 32
A labourer preparing mortar.

PLATE 33
A labourer entering the minaret with a
bucket of mortar.

PLATE **34**
A labourer manning his work station in the minaret stairwell.

PLATE **35**
Regulating the flow of materials at the top of the minaret.

113

and, when genuine, served to strengthen the relations among the labourers. An apparent concern for others might also hide a competitive edge that was in reality more preoccupied with exposing the weakness of others than consoling them. Yemeni men, while asking about another's health, would firmly grasp one of the other man's knees, shaking it slightly for added emphasis to their inquiry. Knees are associated with virility, and their strength signifies sexual potency as the knees are thought to take the most pressure during sexual intercourse. Cracking knees are a sign of age and diminished vigour, and sore knees may indicate too much sex: hypothetically, both ailments can be cured by eating honey.[102]

One young labourer in his late teens, a head of thick knotted curls, the shadow of a moustache on his upper lip, and a small but wiry physique, was particularly ambitious to make his way up the ladder of success. 'Muhammad the Tobacco Chewer', as he became known to me, was an expert at strategically exposing the weaknesses of his fellow workers, securing every opportunity to do so before an audience of *asatiyyah*. Sensing that his lime-light might be waning, he went to extraordinary lengths to redirect attention to himself, hauling enormous stones or heavy loads of materials upon his back, dwarfing him in proportion, and boisterously challenging the other labourers to follow suit. On one occasion while working at the top of the minaret alongside the *thana'a*, he endeavoured to consume the entire contents of a small plastic bag of powdered chewing tobacco in a single go. His rather comic staging of the feat invoked a great deal of laughter, including from himself as he struggled to maintain the powder within his ballooned cheeks. He proceeded to spew wisps of the still dry tobacco before finally spitting it out. Shortly afterwards, dizzy and nauseous, he made his way back to the ground and passed out curled up in a ball on top of a heap of gravel.

For the most part, however, the labourers operated harmoniously as a cohesive unit, executing their co-ordinated tasks in near perfect synchronicity. One of the most effective devices used by the builders to set and sustain the rhythm of the team work was the singing in unison of *hajl*, or work songs, previously mentioned. These songs have uncomplicated structures and repetitive, almost mesmerising, rhythms. One worker led the *hajl* and the remainder responded in chorus after every line, usually with the same reply or with slight variations. What follows are three examples of the many *hajl* which I recorded at the work site.

Hajl One

'Ali waja' lak	'Ali, it comes to you
ya seen[103] *'alayhu*	God's blessing upon him
'Ali waja' lak	'Ali, it comes to you
ya seen 'alayhu	God's blessing upon him
	(repeat first two verses)

abuuk Muhammad	Your father is Muhammad
ya 'Ali	Ya 'Ali
'Ali tasamah	'Ali is named
Haydarah[104]	Haydarah
'Ali takaram	'Ali is honoured
Haydarah	Haydarah
yadharab biseifihu	He struck with his sword
wa huwa 'asharah	and he is ten[105]
min al-yahud	From the Jews
hud al-fajarah	the blasphemous Jews[106]
'Ali waja' laka	'Ali, it comes to you
bin 'Ali	the son of 'Ali
abuuk Muhammad	Your father is Muhammad
ya 'Ali	Ya 'Ali
alay'ak samah	By your name he's named
ya 'Ali	Ya 'Ali
wa abuuk Tariq	And Tariq is your father
ya 'Ali	Ya 'Ali
Muhammad, Muhammad	Muhammad, Muhammad
ya 'Ali	Ya 'Ali
wa abuuk Saleh	And Saleh is your father
ya 'Ali	Ya 'Ali
Hafaz al-mathahab	Memorise the sects[107]
ya 'Ali	Ya 'Ali
min al-maharag	From the east
ya 'Ali	Ya 'Ali
ila al-maghreb	To the west
ya 'Ali	Ya 'Ali
'Ali waja' laka	'Ali, it comes to you
ya 'Ali	Ya 'Ali

Hajl Two

warad waradeeyah[108]	It's coming, it's (really) coming
waradeeyah	it's really coming
wa al-hadeeth 'asheeyah	And the chatting (will be) in the evening
waradeeyah	it's really coming
Haris al-jamalah	The guard of the camel owners
waradeeyah	it's really coming
malahu mad'anee	What's with him, (why) didn't he call me
waradeeyah	it's really coming
aw shay qad da'anee	Or has it happened that he called me
waradeeyah	it's really coming
aw sharag ba'anee	Or was (is) he late deliberately
waradeeyah	it's really coming

115

fi al-'ashee zeeyarah	In the evening (there will be) a visit
waradeeyah	it's really coming
wa ana jeet ajuurahu	And I came to visit him
waradeeyah	it's really coming
ya bahee binuurahu	Your magnificence is by His light
waradeeyah	it's really coming
wa naraka'a bi'ism Allah	And we prostrate in the Name of God
waradeeyah	it's really coming
wa al-rukun al-yamanee	And the Yemeni corner[109]
waradeeyah	it's really coming
wa nibda' bi'ism Allah	And we start in the name of God
waradeeyah	it's really coming
wa 'ihna waradeeyah	And we are (really) coming
waradeeyah	it's really coming
warad fawq Sana'a	It's coming over Sana'a
waradeeyah	it's really coming
wa Allah 'aduuhu heen	I swear (by God) it's still early
waradeeyah	it's really coming
Ahmad, ya Muhammad	Ahmad, ya Muhammad
ya Muhammad	Ya Muhammad
wa Ahmad wast al-qubah	And Ahmad's inside the dome
ya Muhammad	Ya Muhammad
warad waradeeyah	It's coming, it's really coming
waradeeyah	it's really coming
ya warad al-ma'	Ya, the coming of the water
waradeeyah	it's really coming

Hajl Three

Sharag ya Muhammad	It's late ya Muhammad
ya Muhammad	Ya Muhammad
ya bahree al-tahir	O my pure sea
ya Muhammad	Ya Muhammad
ya bahree al-zakhir	O my bountiful sea
ya Muhammad	Ya Muhammad
ya muslimeen qad qad	Ya Muslims he led
ya Muhammad	Ya Muhammad
jama'ah wa yom 'ithnayn	Friday and Monday[110]
ya Muhammad	Ya Muhammad
wa al-Hureeyah wa al-Hur	And the nymphs and the virgins of paradise
ya Muhammad	Ya Muhammad
wa al-bayrag al-manshur	And the spread (open) banner[111]
ya Muhammad	Ya Muhammad
khalayak hamam al-zayn	Stay, doves of beauty[112]
ya Muhammad	Ya Muhammad

wa 'ana 'alayak Salayt	And I for you (the prophet) prayed
ya Muhammad	Ya Muhammad
Salayt ma malayt	I prayed without becoming weary
ya Muhammad	Ya Muhammad
'ala walee al-bayt	For the saint of the house
ya Muhammad	Ya Muhammad
ya zahr ya taahir	O pure flower
ya Muhammad	Ya Muhammad
ya bahree ya zakhir	O bountiful sea
ya Muhammad	Ya Muhammad
ya man samayt 'ismahu	O, whom by his name[113] I've been named
ya Muhammad	Ya Muhammad

Simple lyrics lent themselves to variations and playful modifications which showed off the wit of the song master and provided an element of amusement. Aside from setting a work tempo, these songs also fostered a sense of unity by providing a shared sanctuary for the minds of the workers, away from the considerable physical hardship and monotony of their labour. Team unity is the hallmark of the traditional building trade in terms of its functioning, and is the characteristic which safeguards the hierarchy of power within. Burckhardt writing on Islamic art and unity states that unity is the central theme of Islam, and "in Muslim art, unity is never the result of synthesis of component elements; it exists a priori, and all the particular forms are deduced from it".[114] It is this ideal unity which shaped any newcomer's understanding of the system, the machine, to which he had to submit his own identity as a labourer.

The unity of the labour force was somewhat more pronounced on projects such as the reconstruction of Sana'a's defensive wall in mud *'zabur'* style, which was currently being erected by teams from the northern city of Sadah. Unlike construction at the minaret where each builder may have performed a specific task in the assembly that was spread out along the length of the spiral stair, (sometimes involving more than fifty vertical meters between the first and the last members of the team), the city wall labourers all performed similar tasks working within close proximity to one another. Two sub-teams of six to eight men worked in parallel, passing up balls of a mud and straw mixture to the Master Builder poised on top of the wall. He caught the projectiles, often two at a time, and hurled them with quick precision into place, effectively building up the height of the wall in even, rectangular sections. As the height of the wall increased, they supplemented the number of team members accordingly. The two sub-teams were spaced evenly along a path leading from the piles of fresh mud, and up adjacent ladders to the *usta* on the top. *Hajl* were sung continuously for the entire duration of the workday and played a much more prominent role in regulating the rhythms of the highly repetitive work. The *usta* modified the speed of the entire process by regulating the tempo of the *hajl*. The song and the performance were united as a single piece of machinery: if the tempo of the song faltered, so too did the performance, and vice versa (see plate 36).[115]

Finding one's role within the mechanics of the system and accomplishing tasks was learned by monitoring one's colleagues and mimicking their performance. In explaning the how children come to acquire the habitus of their class or group, Bourdieu correctly notes that "the child immitates not 'models' but other people's actions ... which, in their eyes, express everything that goes to make an accomplished adult".[116] In the initial stages, mimicry was done in a rudimentary fashion whereby the main objective was simply the accomplishment of the task at hand, or arriving at state B from A. At this early stage, when confronted with a new task that the initiate had never previously performed, the task, or transition from state A to B, was conceived of holistically. In other words, the task, as a newly introduced object of knowledge, was conceptualised at a macro level and therefore devoid of a refined definition which would clarify the intermediate steps for a more economic journey from A to B. The journey itself was accomplished by reliance on what the body already *knew*, and this knowledge may, or may not, have resulted in the most effective performance. However, once this first stage, the arriving at B, was grasped conceptually, and by this I also mean bodily, the trainee was able to reposition himself at a different cognitive vantage point, closer to the so called object of knowledge, and begin to assess the microphysics of the task at hand. This equated to closer observation of one's peers, and, more specifically, of *how they are doing*, as opposed to simply of *what they are doing*. Closer observation and continued practice resulted in a more refined mimicry. Again, once this new level of knowledge was thoroughly inculcated as performance, the trainee could zoom in cognitively for yet a closer view, and so forth until the task was mastered and the knowledge reproduced.

A friend in Sana'a recounted to me how when he was a young man, curious to know the world around him, he used to ask his mother repeatedly why he had to pray. She would reply each time with the same answer, saying very simply "because we are Muslims, we must pray to God and give him thanks", and would subsequently send him off with his father to the mosque. Next to his father he made his prostrations, keeping a close look out the corner of his eye to make sure he was keeping time with him, and the whole while feeling slightly unsatisfied with his mother's explanation, but never questioning further. Likewise, the labourer at the minaret did not receive a satisfactory explanation about his role in relation to the global dimension of the project, but, for the most part, nor did he expect one. Without words, he was obliged to have 'faith' in the system of which he was now a part, to subject himself fully to its disciplinary apparatus, embodying the means of an economic performance.

PLATE 36
Re-building the city walls of
Sana'a.

The labourers were frequently audience to the discourse between the Master Builders which included issues of project scheduling, budgets, payments of materials, payment of wages, builder-client relations, dealings with suppliers, and so forth. These subject realms were not open to participation from the young workers, but those who were motivated for success in their vocation listened intently. Part of their complete training as traditional craftsmen required a knowledge about all aspects involved in the building process, including economics and the management of resources. In a sense, this knowledge had to be 'stolen' through 'passive' listening, since the trainee's role, as it was defined by this system, was not to actively participate in the discourse of trade lore. Initially, they may have soaked up only those issues which had a direct bearing on themselves, such as work schedules or wages, and gradually they expanded their interests and their conceptual grasp of more complex topics that were part of the programme weave. However, it was not until they were taken on as an apprentice (i.e. head labourer), working continually in the shadow of the *asatiyyah*, that they would have adequate exposure to these discussions and therefore be capable of formulating the connections between various issues and arrive at a more comprehensive knowledge about their trade.

Many of the labourers, in fact, maintained only a limited knowledge, very often confined to their specific tasks, and even this was sometimes unaccompanied by an understanding of the bearing they had on the overall project. Only a few, when questioned, could offer detailed verbal accounts about the structure and how the actual building was being executed by the *usta*. The latter were normally those who had worked on minarets in the past with the Bayt al-Maswari, and, in particular, those who had displayed a keen sense of motivation to one day become Masters in the trade. However, although they had already achieved a certain level of understanding regarding many of the trade skills through simple observation, labourers, being on the lowest rung, were only permitted to *master* their own tasks and, significantly, self-discipline. To become a master of the trade meant *mastering making*, and before being allowed to *make*, one had to discipline the hand, and thus the mind, that would do the making.

Concluding Remarks

In effect, labouring can be viewed at once as serving both immediate practical requirements on the site, and as a necessary stage in a training process which is ultimately designed to control the reproduction of the trade knowledge. Practically, labourers perform necessary tasks which enable the higher ranked builders to physically build. The training, on the other hand, serves to instil a sense of discipline with regard to the trade ethic and obedience with regard to hierarchies of power and knowledge. Sedimenting discipline and obedience in the actions, behaviour, and attitudes of the young builders serves to lay the 'foundations' for their future as members of the trade, and determines the 'principles of structure' for the formation of that future. Thus the labourer comes to embody the foundations and principles of the system. As in the case of the edifice, setting foundations and determining the principles of structure is the most crucial step in safeguarding the 'structural integrity' of the trade, in the present and over the long term.

In comparison with Islam, the nature of this first stage in the training of a builder is fundamentally the same as *islam*, the first dimension of the Faith. The child is inculcated with a set of practices which lay the foundations for his/her membership in the religious community, and which determine the principles by which the individual will gradually develop an understanding of themselves as Muslims. Both factors are crucial for the reproduction and preservation of the Faith, not just at the level of the individual, but, equally, at the greater level of the community and over time. Like the child learning to fast during Ramadan, the initiate labourer learns to perform tasks through a process of observation and mimicry. Both child and labourer are learning *how to act*, but not yet about *how to understand*. Reaching this next stage is largely incumbent upon their own motivation. For the builder, the deeper understanding of his trade will come once, having transcended the status of labourer and being admitted as an apprentice, he engages in the process of making. In the mean time he must seek fulfilment through the mastering of discipline, subjecting himself entirely to the trade code as set out by the Masters. Nasr writes regarding the production of Islamic art and spirituality that:

> essentially the Divine Law contains instructions for Muslims on *how to act*, not *how to make things*. Its role in art, besides providing the general social background, is in moulding the soul of the artist by imbuing it with certain attitudes and virtues derived from the Qur'an and the prophetic Hadith and Sunnah. But it does not provide guidance for the creation of a sacred art such as that of Islam.[117]

Like the training prescribed by the Divine Laws, the training of a labourer is inherently a training in disciplined comportment. Returning to the opening quotation of this chapter, such a training provides the "recourse to rites whose very nature is to cast a sacred form upon the waves of the ocean of multiplicity", and foster a unity. It is in the hope of this unity that the knowledge of the traditional building trade will transcend not only individuals, but time.

Chapter Three

MAKING IT ABOVE GRADE
Apprenticeship and Learning to 'Make'

First we recognise that the object is one integral thing, then we recognise that it is an organised composite structure, a thing in fact: finally, when the relationship of the parts is exquisite [here meaning, as in Latin exquire, "well searched out", with a pun on the normal meaning of the word], when the parts are adjusted to the special point, we recognise that it is that thing which it is. Its soul, its whatness, leaps to us from the vestment of its appearance. The soul of the commonest object, the structure of which is so adjusted, seems to us radiant. The object achieves its epiphany.

James Joyce[1]

In the preceding chapter I framed my analysis of the Sanaʿani labourer's training within the broader context of education in Yemen, and with the early religious training of a Muslim. By doing so I attempted to demonstrate how certain social and cultural factors inform the manner in which a set of practices and a sense of discipline are inculcated in the young builders. Through comparison with *islam*, the first dimension of the Faith, I concluded that in the course of his initial training on site, the labourer, like the young Muslim was learning *how to act*, but not yet *how to understand*. With practised and acquired aptitudes, and a disciplined motivation imbued with obedience to the existing hierarchies of rank, the builder might transcend his status and be accepted as an apprentice. It is through the process of *making* that the novice eventually acquires an expert knowledge of his trade, and reaches a conceptual state of 'understanding', the equivalent of *iman*, the second dimension of Islam.

This chapter, like the last, is divided into two main sections. Again, the first and shorter of the two continues the description of erecting a minaret. From the stone base, the brick shaft of the minaret is raised from the inside out, without scaffolding, reaching skyward to be ultimately capped by a dome. This middle section constitutes the largest portion of the structure, and is lavished with the greatest consideration from its designer-builders. It is conspicuously decorated with panels of relief brickwork which are later picked out in white gypsum plaster, and its towering presence not only marks the centre of the religious community, but also shows off the skill and craftsmanship of its builders across the skyline of the city. Likewise, the apprentice, also in a middle position between 'lowly' labourer and *usta*, is selected to work alongside the Master Builder, and is indoctrinated with the trade skills and the dispositions of this mentor which will shape both his person and his career.

The second section of this chapter focuses on the training of the building apprentice, with reference to my studies of the Bayt al-Maswari in Sana'a. In the first subsection entitled *Apprentices and Apprenticeship*, I present a general discussion of this training period, and address the issue of there being no distinct lexical category used by the local population (and more particularly the Bayt al-Maswari builders) to define either 'apprentice' or 'apprenticeship'. In the subsection *Earmarked for Usta*, I describe the struggle to become, and remain, a 'top' labourer, drawing on the case study of three young builders in competition for this privileged status of serving the *usta*. I conclude that the hierarchical positioning is fluid below the rank of Master, and proceed in the following subsection, *Slippery Categories*, to explore the culturally determined salience of certain taxonomies and lexical categories over others, and conclude by introducing a modular theory of mind as a means to challenge and expand upon the principals of a theory of language relativity.

The modularity approach to studying knowledge is expanded in the final two subsections in which I make comparative studies of the teaching-learning processes of the Yemeni builders with that of carpenters in Quebec and of architects. Within the modularity framework, I conclude, in support of other cognitive research, that the linguistic-based priority of knowledge subscribed to by language relativists, must be reversed in order to expand the investigation of human cognition to include at the very least image-based, sensorimotor, and emotional domains of knowledge.

I About Building: fashioning from the inside out

In Netherlandish painter Jan van Eyck's (d.1441) depiction of Saint Barbara (see plate 37), beyond the foreground horizon delineated by the draperies of her dress, her tower is being erected in the northern High Gothic style of the artist's period. Behind her, at her left elbow, a congregation of women chat, and a second small crowd of men on the left side of the tower seem to be monitoring the progress of the building. Similar to the construction site of a new minaret in Sana'a, the Gothic tower in this early fifteenth-century painting seems to be a focus of public interest, attractive perhaps for the views from the top, and therefore serving as a local forum for social interaction. Saint Barbara, her back to the tower which will become her

PLATE 37
Jan Van Eyck: Saint Barbara,
1437. In Antwerp, Koninklijk
Museum voor Schone
Kunsten.

prison, looks placidly into the left hand corner of the composition
seemingly unaware of the hive of activity behind her. With her right hand
she turns the pages of the book which rests in her lap, likely the Bible, while
with her left hand she lightly clutches the palm frond, the symbol of
Christian martyrdom. Barbara is recognised as the patron Saint of
quarrymen, as well as of artillerymen, miners and firemen.

An inspection of the Flemish figures engaged in the construction, busied
in their activities around the base and on top of the extending structure,
prompted my immediate comparison with the minaret builders of Sana'a.
Several groups of labourers are portrayed at the ground level preparing
materials, and one figure is shown entering the doorway of the tower,
perhaps transporting tools or materials via an interior stairwell to the top.

Rough stones from several piles around the site are being transported by cart and by hand to a team of stone masons assembled under the shade of a temporary awning. The finished blocks of cut stone are then hoisted with the use of a wooden crane which is positioned on top of the tower, and which extends skyward beyond our view, outside the upper frame of the composition. A figure stands waiting at the edge next to the crane for a stone sent by his fellow labourer below. Likely, he will be responsible for transporting the finished building materials to the team of Master Builders, and I imagine that he may be an apprentice of sorts. The four Master Builders depicted by Van Eyck are engaged in setting the masonry of the outer walls of what appears to be an octagonal planned shaft. Curiously, there are no signs of exterior scaffolding. Rather, like the Sana'ani minarets, the structure is being erected from the inside out.

With the completion of the minaret's stone base as described in the last chapter, the erection of the shaft proceeded in kiln-baked brick, *ajurr* (see plate 38). Construction during all phases of the project, from beginning to end, was executed from the inside out, thereby eliminating any need for external scaffolding. In effect, the structure of the tower physically supported the processes of its own vertical extension. All components of the plan rose in tandem whereby the addition of steps and the rising height of the central column were co-ordinated with the rising height of the walls. In this manner, all building assemblage was executed at the top of the structure, generating growth uniquely in the vertical plane. Variations in plan along the horizontal plane were minimal with the exception of the protruding balcony and the subsequent decrease in the width of the walls from four bricks thick to three. The diameter of the central column and the width of the stairs remained more or less constant throughout.

The staircase, measuring between ninety centimetres and a metre in width, provided the main artery for the flow of men and materials to the top while the minaret was being constructed. Before the advent of electricity and the loudspeaker in Yemen, the *muezzin* was required to climb the minaret five times daily to give the call to prayer from the balcony. Nowadays, the minaret stairs are rarely climbed except to service the loudspeakers or for maintenance of the structure. I was informed by one of the *asatiyyah*, however, that it is necessary to build a solid spiral staircase in order to ensure the tower's structural integrity. The stair behaves as a continuous brace from bottom to top, spanning between the solid, but slender central column and the outside walls, and thus renders the tower more rigid as well as resistant to lateral forces.

The *usta* went on to tell me that many of the modern-style minarets, pointing to the twin minarets of the Martyrs' Mosque outside Bab al-Yemen to exemplify his claim, are not built with staircases, but only ladders, 'sullam min ma'din', and discredited them as fragile. The detectable superiority of his tone signalled the competitive discourse that I frequently

PLATE 38
Construction of the brick tower above the stone base of the al-Ihsan minaret in Madinat al-Asbahi.

witnessed between traditional builders and modern architects and engineers, even though to my knowledge the parties rarely encounter one another directly. In fact the Martyrs' minarets did contain staircases, but they were built of cast-in-place concrete and there appeared to be no contact between them and the open decorative portions of the walls at the base of the shafts. Rather they were cantilevered off the central reinforced concrete column and were therefore not serving to brace the towers.

Individual steps were constructed riser by riser as the height of the external walls was raised in brick. Beams, each comprising a pair of thick tamarisk branches, were spanned between the central column and the external wall. On average seven of these beams were positioned per revolution of the spiral stair at equidistant intervals, ascending vertically in correspondence to the required slope of the staircase. Between these beams, longer, thinner, more flexible branches of tamarisk, known as *al-sab'*, 'fingers', were laid closely packed together, bent in conformity with the curved space between the two beams. This wooden superstructure supported sixteen stone stair treds per three-hundred-and-sixty degree rotation of the wind. The stair treads of roughly hewn stone were set in mortar, and the space between them and the *sab'* supports was filled in with mortar and rubble.

In the lower half of the shaft, tamarisk sticks were also laid flat within the walls after the completion of every fifteen to twenty courses of brick in order to strengthen the structure's lateral resistance. The sticks were cut to short lengths and arranged radially pointing outwards, set one next to the other between an exterior and interior facing of half-bricks. In his study of traditional architecture in Saudi Arabia, King also notes that wood courses were used in conjunction with the coral masonry on the Red Sea coast, as well as in the stone buildings of the Hijaz. He proposes that it is "an ancient technique indigenous to the Arabian peninsula" with examples dating to between the fifth and third centuries BC.[2] Tamarisk, in Arabic *athl*, is the wood most commonly used in construction in Sana'a, growing abundantly in several wadis outside the city and in those fingering their way down from the highlands to the Tihama Plain.[3] Several literary sources, including *The City of Divine and Earthly Joys*,[4] allude to the employment of star calendars, *ma'alim al-zira'a*, for judging the most appropriate season for harvesting woods so as to avoid worm infestations, though, according to his recent studies, Varisco comments that only a few contemporary farmers admitted to observing the star sequence anymore.[5] "Why bother" he continues, "when there is a calendar on the wall and the date flashes from an inexpensive wrist watch?".[6] Jamal al-Din 'Ali's eighteenth-century account informs us that the:

> tamarisks planted for use by former generations ... have grown into shady thickets inhabited by wild animals and flourishing with such extraordinary fertility that they equal in height the loftiest cypresses. The people of these towns (around Sana'a) rely mostly on these tamarisk for building and (other) work, and to make their vine trellises.[7]

The tamarisk trees still provide an important source of building materials, and the trellises to which the author refers abound in the present-day vineyards of ar-Rawdha, Wadi al-Sirr, and Wadi Dhahr outside the city.

The consistent diameter of the interior face of the brick wall wrapping the radial staircase, and the dimensions of the transforming geometries of the wall's exterior face and projecting balcony, were all measured from the centre point of the minaret's circular plan, marked by the metal pipe that rises through the entire height of the structure (see figures 6 and 7). This hollow metal pipe, approximately four centimetres in diameter, is called *qasabah* in Arabic, and is also the word used to refer to the human trachea. It is manufactured in manageable lengths of a few meters, and threaded at the ends so that individual sections may be screwed together. Consequently,

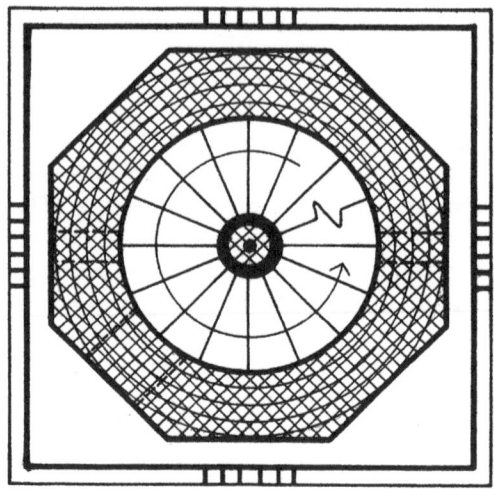

FIGURE 6 Plan of the octagonal brick shaft rising above the square base.

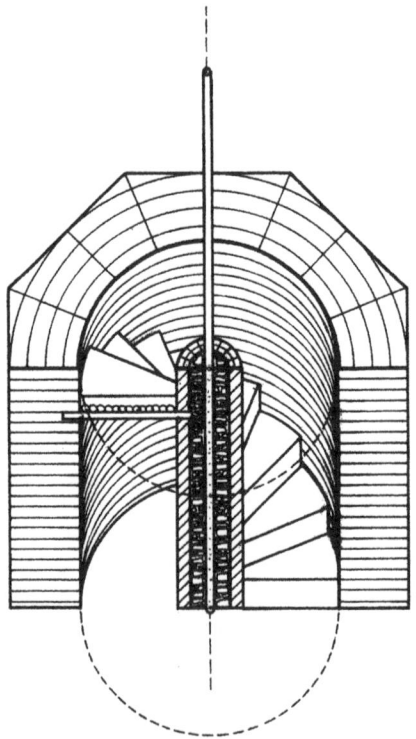

FIGURE 7 Section through the octagonal brick shaft showing the central metal pipe, or *qasabah*.

PLATE 39 Building the stone column around the central metal post, or *qasabah*.

the pipe may be conveniently extended in co-ordination with the rising height of the walls. This pipe was maintained in a perfectly vertical position, and rigidly encased within the central stone column of the stairwell (see plate 39). Firm warnings were issued to all newcomers that the *qasabah* was not to be tampered with under any circumstances. Nor was it to be clutched by the apprentice who precariously straddled above the staircase, one foot planted on top of the wall and the other on top of the column, while he distributed materials to the *asatiyyah*.

Interestingly, the column itself is called the *qutub* by the builders, which appropriately translates to axis or axle as all dimensions are measured and verified radially from the metal post at its centre. In contrast, the general word used to describe a column which supports a roof or an arch is *'amuud* (plural *a'midah*), derived from the verb *'amada*, 'to support'. Rather than supporting a roof or structure above it, the central column of the minaret ties the various structural components of the tower together laterally. A similar role is played by the square stone pillar at the core of the stairway in a Sana'ani tower house, which is said to shoulder the integrity of the edifice – if it fails, so too does the house.

As discussed in the previous chapter, all geometries of the minaret plan were measured from the centre using a nylon rope, *habl*, attached to the metal post and marked along its length with knots or nails representing the dimensions of the plan's geometries. For example, the geometries of the sixteen flutes expressed on the lower elevations of the brick shaft were

129

PLATE 40 Ahmad al-Maswari, a Master Builder, carving bricks.

constructed using two points along the rope, the first corresponding to the fillets which frame each flute, and the second to the trough of the flute itself. Half bricks were used to fill in the remainder of the curve by eye, following the pattern as initially set by the head of the Master Builders. A third knot in the rope marked the diameter of the interior stair wall. Verticality was verified continually on either side of the rising brick wall using a plumb-line. The building proceeded in the counter-clockwise direction, like the ascent of the stairs, and perhaps symbolically like the direction of the *Hajaj* (circumorbital ring around the *Ka'ba* in Mecca), spiralling heavenward around the 'axle'. This, however, was never explicitly stated by the builders themselves, but rather suggested by other Sana'anis that I spoke with. The metal post of the axle, *qasabah*, invariably extended beyond the height of the current stages of construction, and the Master Builders often draped their tattered sports jackets over the top, flagging their presence on site to all below.

Crucial elements in the geometry, such as the eight corners of an octagonal section of the tower, were set out by the head builder who began at one point on the horizontal plan and continued, almost without exception, to complete the circle in a counter-clockwise direction. He constructed each corners of the octagonal plan with a single square *ajurr* brick by adjusting it with his adze, *mitraqah*, to accommodate the necessary one-hundred-and-thirty-five degree obtuse angle between faces (see plate 40). His assistant passed him wetted bricks which had been soaked in a large metal basin supplied with water via a rubber hose that spiralled around the column, following the stairs to the top. The kiln-baked clay bricks were highly absorbent and thus easier to carve, as well as amenable to better cohesion with the clay and lime mortar. Squatting on top of the wall, the *usta* clasped the brick firmly in one hand, standing it in a soldier-course formation and orienting its length perpendicular in relation to his body. With his *mitraqah* in the other hand, he lopped off the corner nearest him usually with a single clean blow, and produced a nearly perfect forty-five degree notch. He then laid the brick in position at the intended corner in order to verify the fit of the angle by eye. If necessary, he made final adjustments to the brick, then smoothed the cut face with the edge of his adze before setting it in mortar.

After setting each corner brick, the builder mechanically verified the verticality by dropping his plumb-line down the face of the wall. Two courses were laid before the *usta* proceeded to the next corner. Once two corners had been constructed, a thin white cord was fastened tautly between them, running over their outer faces and demarcating the outer boundary of the linear wall to be constructed between them. A second *usta*, normally of lower rank, followed in counter-clockwise direction, filling in the outer face of the wall with alternating courses of full and half bricks to secure a solid bond, and possibly advancing a decorative pattern initiated by the head *usta*. A third *usta* was responsible for simultaneously laying the brick courses of the plain interior wall face, also in alternating courses of full and half bricks, and the *thana'a*, below the level of *usta* and derived from *ath-thaani*, 'the second', was accountable for infilling the wall space between the faces (see plate 41). Because the geometry of the round wall was not conducive to the perfect fit of rectilinear building units, infilling was effected with an odd assortment of triangular and trapezoidal brick shapes and pieces of rubble set in mortar, with the principal aim being to create strong bonds between courses and to eliminate any air pockets. The extent of the visibility of the craftsmanship was, for the most part, directly correlated with the builders' hierarchy: the exterior facade was executed by the senior *asatiyyah*; the rarely viewed interior wall face was completed by the lower ranking ones; and the hidden workmanship of infilling the wall was performed by the *thana'a*. This hierarchy was also observed by Varanda in his study of traditional building in Yemen in which he states that the *usta* is responsible for laying the outer face of the building, the *thana'a* sets the course of masonry on the inner face, and the *rassah* fills in the wall between the two surfaces.[8] Unlike Varanda's straight forward account, however, these roles were not strictly assigned to any particular designated team members at the minaret site, and it was not uncommon for the head *usta* to 'fill in the walls', or for the *thana'a* to occasionally participate in setting the interior, and less commonly the exterior, faces of brickwork.

The bricks produced for all construction were of two standard dimensions, a whole brick measuring about sixteen centimetres square by

PLATE 41
The *thana'a* infilling the brick walls.

five centimetres thick, and a half brick, called an *ansaf* (derived from the verb *nassafa*, meaning 'to halve'), measuring sixteen by eight centimetres, also with a thickness of five centimetres. In 1997 full bricks were sold by the producer at eight riyals per unit and half bricks for five riyals, approximately six-and-a-half cents and four cents US respectively. Amin al-Mantasar, a brick producer with kilns on the outskirts of Sana'a, felt that it was an honour to be the sole supplier for the Bayt al-Maswari projects, asserting that they are the best House of Builders in Sana'a. Al-Mantasar's circular, hive-like kilns, dug deep into the earth, were capable of baking up to thirty-five-thousand bricks at a time. The clay used by his production unit was quarried north of the city, beyond the international airport, and the bricks were cast in wooden moulds and left to dry in the sun for several days. Afterwards, they were placed in the wood-fired kiln for a period of up to two weeks. The fires were maintained for half that period, and the bricks were left inside to cool for the latter half (see plate 42).[9] Not surprisingly, the quality of individual bricks varied considerably depending on the

PLATE 42 An operational brick kiln south of Sana'a.

consistency of their composition and on their position within the kiln. Bricks of a consistent red-earth colour, as opposed to those with dark or black splotches, were deemed the most suitable for the exterior wall facing and could be carved with ease for decorative purposes.

The inspiration for the decorative brickwork on the al-Maswari minarets was derived from the abundance and variation of decorative patterns on Sana'ani houses and the minarets of the old city, a prominent hallmark of the region's architectural character. It was normally the head *usta* who 'designed' the decoration, *naqsh*, usually when stimulated by the effects of his *qat* chewing, but the other *asatiyyah* also indulged in this creative activity and made occasional contributions to the final configuration. Each of their minarets displayed a unique decorative pattern achieved by subtle variations and manipulation of the limited palette of highly geometric motifs. They firmly maintained that there was no symbolic significance to the patterns they chose, but rather they were simply a part of the aesthetic heritage of their city.[10] Because the bricks produced were of a standard rectilinear dimension, as well as being fairly uniform in their colour, decoration was almost exclusively reliant upon three-dimensional relief arrangements.[11] A fixed number of carved brick types had been developed which could be easily shaped on site with the chisel-like-adze, *mitraqah*, and which facilitated permutations of the raised geometric arrangements. The carving of bricks will be discussed in greater detail later on in this chapter.

In comparison with the builder's eyed judgement in determining the quantity and angle to be removed from an individual brick with his adze, his judgement of spatial relations of greater magnitude was generally reliant on more precise numerical systems of measurement. For instance, the gauging of the vertical progress of the structure was qualified in metres as measured with a tape, or alternatively by the number of brick courses, where sixteen bricks was known to equate roughly to one metre. On the other hand, at the project's scale of greatest magnitude, when the parts of the minaret were to be adjusted in relation to the whole, the proportions of the overall structure were judged without the use of mathematical ratios. The experienced senior *asatiyyah* judged the aesthetic value of the minaret's proportions 'by eye' and determined whether the height, which had been stipulated by the patron's budget, should be modified during the course of construction in order for the minaret-as-object to (in the words of Joyce) "achieve its epiphany".

As the walls rose, the outermost surface of the brick was rubbed smooth by the apprentice who applied daubs of wet, runny mortar to the surface and scrubbed it into the brick facing with another brick which he held like a shoe-shine brush. In effect, the patina of the brick was removed by the abrasive rubbing action, and replaced by a thin, even layer of earth-red plaster produced by the mixing of the mortar and the pulverised brick. The resultant outer surface of the structure was smooth and uniform in colour, rendering the seams between individual bricks nearly undetectable. However, the long-term consequences of removing the kiln-baked patina are highly questionable, and will likely induce a premature erosion of the brick, not dissimilar to the aftermath of sand-blasting Gothic cathedral masonry in the effort to 'clean' it. One Yemeni affiliated with the General Organisation for the Preservation of Historic Cities in Yemen (GOPHCY) dubbed this widespread recent fashion as "The Cancer of Sanaʿa".

Throughout the construction, narrow slit windows were introduced in the minaret walls at every half turn of the stairs, alternately aligning with the east and west cardinal directions, and supplying natural light to the passage.[12] After nine complete rotations of the staircase in the minaret of the ʿAddil Mosque, or one-hundred-and-forty-four stairs from the ground level, an exit was provided onto the calling platform. The exit door was low and shoulder width, and was spanned by a corbelled arch and a lintel of tamarisk boughs, reminiscent of the corbelled arches used in the Mayan architecture of the Yucatan. The platform itself was slender in width, about seventy-five centimetres at its widest point, ringing the shaft, and protected by a low brick wall. The platform was assembled by progressively cantilevering the courses of bricks out over each other, setting the individual square bricks diagonally in a tooth-like arrangement. A strong quicklime mortar (in Arabic *jiss* and pronounced *guss* on the site) was used exclusively at this stage to bond the courses of the projecting platform. The subsequent railing-wall with its eighteen decorative panels would only be constructed at the end of the project, as it served no pressing structural purpose for the continuation of the tower.

In the case of the 'Addil minaret, construction of the octagonal shaft above the balcony proceeded with decorative panels of brickwork on all eight faces. The wall thickness in this upper portion was now reduced from four bricks to three bricks (approximately fifty centimetres) at the narrowest points which lie between the curve of the inner wall and the centre-points of the exterior planar faces. Another thirty-one stairs, totalling one-hundred-and-seventy-five, finally led to the uppermost level. This interior platform was surrounded by eight large corbelled-arch openings in which the sentinel presence of loud speakers replaced the *muezzin*. The stairs and the central column terminated at this point in the structure, but the *qasabah*, the metal post at the centre of the "axle", continued its vertical extension and ultimately pierced the dome some eight metres above, and secured the *hilal* (crescent moon) above the minaret, a marker of the Islamic faith. These final stages in the construction process will be discussed further in chapter four.

II Apprenticeship and Making

Apprentices and Apprenticeship: a general introduction

'Apprentice' may be translated in Arabic as *ghulaam al-mumahan*, literally 'vocational servant', and 'apprenticeship' as *al-tamahhan*. However, according to 'Abdulwahhab al-Sayrafi, a member of an established family of Sana'ani carpenters who have trained others in their craft, there is no distinct local term or clear definition for an apprentice. Sergeant records that the labourer on the building site working alongside the *usta* and passing materials onto him is called the *munawil*,[13] and Varanda uses the term *tilmidh*,[14] but I did not encounter either of these terms during my experiences at several work sites. In fact, despite great efforts during the first few months to ascertain a clear category for the labourer working alongside the *usta*, none surfaced. I would periodically interject *muta'allim*, which officially translates to 'apprentice', and *tamreen*, which can mean 'practice or training' as well as 'a period of apprenticeship', in an attempt to coax a term from the Master Builders. They clearly understood the Arabic references I used, but did not employ them themselves in the context of their own professional practice. When turning to my fellow labourers to supply a title for this position, the response was simply *shaqi*, 'a labourer', like the rest of them. Once, while persisting in this matter, I was furnished with the term *sabiy al-usta* which translates literally as 'lad of the *usta*'. The hesitancy on the part of my fellow builders in issuing this alternative response clearly indicated that it was aimed at satisfying my, perhaps misdirected, inquiry more so than being a popularly employed term.

Although the distinct status of that builder which I recognised as an 'apprentice' was plainly acknowledged by all, there was seemingly no special terminology to designate it. This was not entirely surprising since, as Bloch notes, it is plain that there need be no inevitable connexion between concepts and words, as has been demonstrated by studies of conceptual thinking in pre-linguistic children,[15] and by cognitive research premised on modularity theories of mind.[16] Nevertheless, the absence of the term became increasingly intrigueing as its relation to the building site politics was revealed. I will discuss these related issues in greater detail in the course of this chapter.

The categorical absence of a precise term in the local use of the language also prompted consideration as to whether one should or should not use a distinct English term in describing that particular agent engaged in that particular role, and assuming that particular status. For the purpose of the study it might have been more appropriate to abandon the notion of 'apprentice' altogether, and replace it with a more transparent translation from the Arabic terminology used by the builders to explicate the role: "*al-shagi alathi yishtaghal bi-janb al-usta*", literally "the labourer who works alongside the *usta*". However, aside from being cumbersome, I determined that the special role undertaken by this worker does indeed match the role defined by the English term 'apprentice'. The categorical absence had nothing to do with whether or not there was a general recognition of the distinctiveness of the role, which decidedly there was, than an actual lack of definitiveness concerning *who* might be filling it at any given time. Therefore, for the purposes of my study, I have maintained both the terms 'apprentice' and 'apprenticeship', and borrow Michael Coy's definition to introduce a general discussion on this special teaching-learning relationship:

> Apprenticeship is a complex and multi-faceted concept. It clearly involves education, social relations, and economics, and it suggests an ideology of life and work associated with a special role. Apprenticeship involves at least two persons and probably more than two. The two principals are a person possessing specialised skills and a person who wishes to acquire and develop those skills for him/herself. Apprenticeship thus consists of a social relationship. And, inasmuch as the specialised skill sought by the apprentice are often, perhaps always, valuable, apprenticeship has an economic dimension.[17]

Generally, a system of apprenticeship not only serves to reproduce knowledge about a craft, but also acts as a valve to regulate the numbers permitted access to this realm of specialised knowledge.[18] Normally, the first encounter between a young apprentice and the trade would be when they are accepted into the shop of a Master, and employed to effect menial tasks. Depending on the nature of the trade and the context, this first step may be accompanied by some form of initiation ritual as described by Ibish in his discussion of craft guilds in the Islamic Middle East.[19] In some places it is probable that no salary will be paid for the first few years, but the apprentices may be given food and lodging while learning the craft and being socialised into their related community role,[20] or in some cases the apprentices may even be required to pay for their training.[21] In these circumstances, when the apprentice reaches adulthood they may receive a nominal wage. As described in the previous chapter, the apprentice at the building site in Sana'a was paid a daily wage throughout the entire period of his labouring and was expected to finance his own board. Despite the variations of economic arrangements between Master and apprentice, it is a universal characteristic of this system that the novice is not permitted to open his/her own practice or accept clients until he/she is deemed to be fully qualified by their Master.

As I will explore more fully in the following chapter, the power of a Master Craftsperson is vested in his/her secret knowledge of the trade, which can be expected to be shared by others of equal stature in the particular society. This knowledge had been inculcated throughout the process of apprenticeship, and great care is taken to ensure that it is not divulged to the uninitiated or to potentially harmful competitors. The restricted confines placed on the discourse produce a hierarchy of power relations between craftsperson and client, or between those who control knowledge and those without, and equally serves as a cohesive force amongst trade members. This constitutes a major component in the construction of a distinct identity which elevates the master craftsperson outside the bounds of normal social relations and informs their corresponding behaviour.[22]

This form of specialised discourse entrenched in its own historicity and striving to conserve underlying cultural models is what Pascal Boyer terms 'traditional discourse'. One of the key aspects of such discourse is that it emphasises specific events and contexts rather than theoretical inferences, therefore it is "based on representations which are essentially different from common stereotypes".[23] Traditional statements made in the customised format of the discourse by those 'customised', or initiated, persons, like the Master Craftsperson, are deemed to have truth value. The judgement of truth is not necessarily based on the content of the speech act, but rather on the utterance as an event being produced by the customised person who embodies the supposed historicity and expert knowledge, in this case, of their trade. "The acquisition of expert discourse implies, not the acquisition of new general representations about the entities designated, but the memorisation of ostensive presentations of these entities and properties",[24] and the communication of these is not restricted to verbalisation, but may well include, or consist entirely of, expressive actions or the use of symbolically imbued objects.[25] The intention of traditional discourse as a form of communication is not the transmission of particular information, but the deliberate and adroit modification of the receiver's conceptual representations and metarepresentations of the world,[26] thereby confirming their own hierarchical position within the social structure.

It then seems paradoxical that the aim of an apprenticeship system is to instruct others in the guarded knowledge of the discourse, when such a proliferation would seem to undermine the power associated with restriction. This difficulty would seem to necessitate the inculcation of social responsibility as a crucial part of the apprentice's education.[27] I will argue in this chapter, however, that the nature of the builder's expert discourse is not an objectified form of knowledge, and cannot be passed off in any form other than the apprenticeship. Therefore, social responsibility must involve more than not divulging secrets since, as a non-objectified expertise, it cannot be 'talked about'. It would seem from many of the sources on this subject that specialised knowledge is not simply handed over to the novice, but rather the apprentice must possess a vital dose of their own initiative in order to acquire it. Michael Herzfeld, studying apprenticeship in Greece, has referred to such acquisition as 'stealing', but suggests that artisans actually encourage this behaviour. Those pupils adept at this sort of devious acquisition of knowledge may well be demonstrating their potential as future partners.[28] As discussed in the previous chapter, those builders who transcended the lower ranks of labouring to become apprentices had to demonstrate an aptitude for the tasks, a motivation to both work hard and learn, and the aspiration to one day become Master of the craft.

Since much of the learning process involves little or no verbal communication, the apprentice must rely on his/her eyes, ears, and sense of touch to incorporate their Master's skill into the reproduction of bodily representations of knowledge.[29] Questions are rarely posed, as they would likely be construed as a challenge to the Master's authority, and also, as I stated above, because of the nature of this form of knowledge, it cannot be 'talked about' anyway. This produces a very particular sort of learning relationship that relies on careful attention and cunning mimicry.[30] Herzfeld notes that most verbal communication which does occur is directed toward the apprentice in the form of aggressive teasing and provocation, which he suggests is meant "to goad the youths into showing their mettle". He continues that "the oppression of apprentices by their masters ... reproduces large inequalities ... [which] can be reduced and eventually negated through acts of 'theft'", or the appropriation of expert knowledge.[31] The seemingly belligerent nature of such interaction makes most forms of apprenticeship incompatible with kinship, and particularly agnatic, ties. This position is addressed by Goody who writes that "it is not accidental that apprentices are not learning from their fathers, since the social structural conditions for apprenticeship – increasing complexity of the division of labour, and entry of domestic production into the market – lead youths to seek to learn occupations different from those of their fathers".[32]

My own findings during my studies in Sana'a were that it was actually considered desirable to train one's family members in the lore of the craft, and to maintain the affiliation of one's family name with the associated prestige of being a trade Master for as many generations as possible. Nevertheless, when I first began my field investigations and sought out families of builders whom I had read about in texts published in the 1970s and 80s, all but a small handful had retired from the business, their children now in other occupations, and the legacy of their family names replaced by new ones. I was told repeatedly that this was the typical fate of the great building families, that by the fourth generation their glory had expired and their progeny dispersed into other livelihoods. The echoes of Ibn Khaldun were uncanny. He wrote in Muqaddimah:

> [Prestige] reaches its end in a single family within four successive generations ... the builder of the glory ... knows what it cost him to do the work, and he keeps the qualities that created his glory and made it last. The son who comes after him had personal contact with his father and thus learnt those things from him ... the third generation must be content with imitation and in particular, with reliance upon tradition. This member is inferior to him of the second generation, in as much as a person who relies [blindly] upon tradition is inferior to a person who exercises independent judgement.[33]

In fact, this seems to parallel the course of the Bayt al-Maswari for whom I worked during the course of my field study. An uninterested member of the family's fourth generation of builders worked sporadically between academic studies, and a young member of the third generation showed little promise in becoming a serious candidate to carry the prestige of the family name. The head *usta* had already taken on an unrelated member into the circle of Masters, and another was being trained. In the following section I will present personal profiles of several of the higher ranking labourers who were either apprentices, or contenders for the position, and discuss their qualities as potential *usta* in light of the training process discussed in the previous chapter, and in consideration of the needs and aspirations of the current *asatiyyah*.

Earmarked for Usta

I slept very little the night before my first day on the building site. I was electrified by contrasting currents of emotion: I was anxious to work on the top of a minaret, to play a part in adding fabric to a city I adored, and to be commencing what I hoped would be invigorating fieldwork. I was also nervous. What would they make of me? A Western Christian who spoke in a somewhat stiff Modern Standard Arabic, and not in the Sana'ani *lahja* (dialect). And what about the work? How dangerous would it be? Would I stumble on the rim of the minaret wall and plummet thirty-five metres to my death? A quick, but dramatic end to my life. Would I be capable of performing the work tasks, or forced to come to the ego-deflating conclusion that despite all my past physical training I would never keep up with those 'small but wiry' local builders. The next morning was to be the start of my discovery that I would confront all of these over the next year (except, thankfully, for the dramatic plummeting), as well as countless other experiences, both good and bad, that I could never have foreseen at this early stage.

I arrived at 6:30 am as instructed by Ahmad, the younger of the two elder al-Maswari brothers. There was no one around and the site was locked. An elderly man with a brightly coloured head wrap and cane hobbled slowly out of the mosque from his morning prayers. I asked him about the builders. He was hard of hearing. I shouted. "Oh them, they won't be around for a while yet. Maybe an hour, *in sha' Allah*". Was I being tested? Even the old man knew they don't start at this time. Or were times from cheap plastic digital watches being bantered about flippantly. From my past experiences in this country, I decided against the trial hypothesis and vouched for the latter explanation. Anxiously, I returned forty-five minutes later. Still most of the labourers hadn't come, but from the street below I had seen several workers at the top of the minaret. The door at the base was open wide and I began what was to be one of many ascents up the already nearly forty-metre tower. Ahmad al-Maswari was squatting on the narrow rim at the top of the wall, dragging slowly on his cigarette, drinking in a mixture of smoke and the fresh morning air, and staring out vacantly over the sprawling city to the west. The morning sunlight climbed along his back. The chilled air was fresh and there was a light breeze.

He turned slowly to greet me, revealing what I thought was a twist of surprise on his face. Whether the surprise was vested with positive or negative feelings, I was unsure. I suspect that he never actually expected me to turn up to work. The three others there, quietly busied in their tasks, individually mumbled greetings to me then said no more. This was not the warm reception I had received the first two times that I had come up the minaret on visits. Perhaps it was too early in the day for social etiquette. Feeling awkward and not knowing when or how I would be fitted into the process, I took a place on top of the wall and squatted down next to Ahmad to observe. One of the other three, whom I wrongly estimated to be in his mid to late thirties, was absorbed in setting bricks and was evidently a key builder. He was sharply dressed in comparison to any other builder I had seen, sporting a crisp white *zannah* (long, ankle-length cotton shirt), a new olive-green wool vest sweater over the top, his *jambiyyah* hanging from a broad embroidered belt, and polished ankle-cut black leather boots that matched his seemingly severe disposition.

In time I discovered that his name was 'Abdullah as-Samawi, not an al-Maswari and with no blood connection to that family. He was the youngest of four *asatiyyah* working on the site and allegedly in his late twenties, though he appeared to me much older. Early on I mistakenly believed that he was in charge of the project. This was mainly due to his stern manner and because the elder al-Maswari brothers would jest that he was the head Master, *al-usta al-kabir*. I was unaware at the time of the potentially ironic edge to their humour. The pecking-order among the *asatiyyah* during my study period placed Muhammad al-Maswari in the top rank, his younger brother Ahmad as second-in-command, Muhammad's son Majid came next, followed closely by 'Abdullah as-Samawi. 'Abdullah came from the region of Dhamar, and from the same village as the majority of the labourers. He began working for the al-Maswaris while still in his teens and was eventually taken on as an apprentice and was made an *usta*. At that time the sons of Muhammad and Ahmad were neither old enough for, nor demonstrated any interest in, this line of work. Since 'Ali-Said, father of Muhammad and Ahmad, came to Sana'a as a building labourer from Maswar in search of an apprenticeship, the family has become increasingly urbanised and many of the younger generation are pursuing professions related to their academic studies.

Contrary to 'Abdullah being the Head Master, his position was probably the most vulnerable as the youngest, most recent, and non-family member of the team. He likely interpreted my strange and sudden presence as potentially threatening to his position, which might explain his initially cautious and somewhat icy attitude toward me. Once he realised that my stay in Yemen was limited, that I was actually studying something other than trying to become an *usta* myself, and that I was perfectly content to be a 'lowly' labourer, his relations with me warmed and we became friends. He was the only builder who made drawings for me as gifts; simple line drawings of minaret elevations made with a straight-edge in either blue and red ink or coloured pencil. Except for a brief (and potentially disastrous) period during which Ahmad al-Maswari suspected that I might be a spy and tried to discourage my presence at the top of the minaret, my relations on site developed positively. I worked hard and proved to be dependable; the workers and I very importantly shared a mildly sarcastic sense of humour; we spent our lunch times together and occasionally chewed qat on Fridays at my home in the Old City or at their shared accommodation; and my presence on site provided a source of entertainment. The number of questions I posed to them was in tandem with the things they were curious to know about me, about Canada and the West, and about the truths of pop star Michael Jackson. Over the course of the year I formed close and personal relations with several of my co-workers, and friendly relations with all of the team members but one, Saat.

Saat was the head labourer and sole apprentice when I first arrived at the 'Addil minaret (see plate 43). Along with 'Abdullah as-Samawi, he was one of the three working at the top on the first morning when I squatted down next to Ahmad, waiting to be directed in a task. The coolness in his initial greeting never really changed. I quickly discovered that his stand-offish behaviour was not directed solely at me, but toward all of the labourers except one young worker from his village whose sister he was engaged to marry. Through 'Abdullah as-Samawi, Saat came to work for the al-Maswaris in Sana'a as a young labourer. After nearly five years he was being slowly promoted, now working as an apprentice under the tutelage of all four Masters, and very guarded about his position. When asked, he thought that he might be seventeen years old, but his fellow villagers said that he must be closer to twenty after comparing his age with their own approximated ages. Only the urbanised al-Maswaris gave their ages with any degree of certainty. During morning tea break and lunch Saat rarely engaged in the conversations of his fellow workers and mostly set himself apart from the general social dynamics of the group. There was a strange power to his quietness and periodic flash of madness in his otherwise vacant eyes that seemed to stare right through anyone speaking with him. While working he would occasionally, and inexplicably, explode with rage, directing his anger by throwing building materials at the poor labourers unfortunate enough to be working near him at that moment. At times he sought a paternal-like affection from the elder al-Maswari brothers, awkwardly pressing up against them for an embrace. He impressed me as rather dim-witted and unquestioningly subservient to the often abusive demands of the *asatiyyah*, but also calculated and competitive with his fellow labourers, in all making him a good candidate for the position.

The 'Addil minaret was completed a few weeks before Ramadan, and a new commission was accepted to build the brick walls of what was essentially a concrete-framed house. The house was one of several belonging to the brother of the President of the Country, and was located in the sensitive tribal area of Sanhan, west of Sana'a. For security reasons I was forbidden to work out there. Following the completion of this short work, the al-Maswaris postponed the commencement of any further projects until after the holidays. 'Abdullah as-Samawi and the labourers returned to their village and during this time, Saat got married. In the interim, I decided to labour for another team of builders who were working

PLATE 43
Saat, the apprentice.

throughout Ramadan on a project with a pressing deadline. This provided me with the opportunity to work on another building type, under different *asatiyyah*, and to make comparisons with my experiences to date. The project-patron was a peculiar, self-made Yemeni who, despite being clearly infatuated with all things Western, was devoted to traditional Sanaʿani architecture as attested by his carefully restored tower house (with all the technological fittings and modern conveniences of an American suburban bungalow). He had recently acquired an existing two-storey house in the Bir al-ʿAzab quarter, and was adding floors in order to convert the building to a specialised school. As I knew the patron personally, I was readily accepted as a labourer by the Master Builders he employed.

As at the minaret site, there was a hierarchy among the three unrelated *asatiyyah*, led by the slightly elderly *Usta* Naji al-Khowbani who donned a bright orange head wrap reflecting his delightful but resolute personality. Unfortunately for him, there are many jokes about the place he comes from. Khowban, like Dhamar, is a rural region that supplies many unskilled labourers for short-term contracts in the capital. *Ya Khowbani* popularly translates to "Hey Stupid!". Below the rank of the three Masters were three apprentices from Yarim whose individual positions corresponded with their ages. At that point, the most senior had been working for ten years under al-Khowbani, the first five of which he had been merely a labourer. The other two had both worked for other builders prior to coming to this *usta*, and one bragged that it had taken him only one year to be granted the responsibilities associated with apprenticeship. As I questioned them between fetching buckets of mortar, each listened intently to the other's responses. They were highly competitive with each other, often bickering amongst themselves, and anti-social with the other labourers. Like Saat, the two more junior of the three were decidedly irritable and were constantly suspicious of my motivations for working. The eldest, perhaps being in the more secure position, was quite amiable toward me, but unwaveringly authoritative with his underlings.

In contrast to Saat's role at the minaret, the apprentices at the school were given much greater autonomy and were somewhat less subject to the continuous scrutiny of the *usta*. This difference was due in large part to the different natures of the two projects. The construction of the school, like a house, involved a greater variety of tasks that could be executed simultaneously, and provided a much larger area for the builders to work separately from one another; whereas the minaret involved essentially the vertical extension of the structure in a confined and precarious work space at the top. Also, because of the height factor, the minaret involved both greater technical precision and a tighter control over the complete building process than a more conventional edifice. At the school, the apprentices were regularly assigned the mundane tasks of completing the unfinished work of the *asatiyyah*, usually one step behind them, filling in the gaps between floor joists on top of interior partition walls with *ajurr* brick, or spanning the *asabi'* sticks between joists, providing a surface on top of which the mud for the floors is piled. I will discuss how trade knowledge is inculcated through the performance of these tasks in the coming sections of this chapter. What is important to note here is that, like Saat, the apprentices bore the brunt of the *usta*'s abuse.

Verbal abuse and reprimands were conveyed in one direction – from the top of the hierarchy to the bottom: the *usta* reprimanded the apprentice(s) and both reprimanded the unskilled labourers. Curses and derogatory remarks were the most common form of communication from 'teacher' to 'learner' either in the assignment of tasks or in the correction of ones in progress. As opposed to explanation, this form of abuse served as a potent disciplinary tool, effectively reinforcing the existing hierarchy amongst the builders. It is an implicit component in the training of the 'docile body' which was useful to the traditional building system, to borrow from Foucault,[34] and as discussed in the previous chapter. I will argue later that skill-related knowledge was actually acquired through the repeated mimicry of a sensorial perceived example, and not through any form of verbal explanation. Much of the verbal communication between ranks on the work site was charged with aggressive intonation, not speech content.[35] While this style of discourse continued to be accepted by the receivers, it served to reinvest authority in the speaker who controlled the expert discourse, and likewise protected the system (which was actually comprised of very limited expert knowledge) from being questioned and challenged in a critical manner. Knowledge, and thus power, remained tightly controlled by a select group as opposed to democratically accessible.

This ability to dish-out abuse was a particularly important skill for the apprentice to acquire if they were to maintain their precious position amongst the labourers. Their lack of a clearly defined title, as discussed above, situated them in a somewhat liminal position between lowly labourer and Master. Therefore the natural establishment of a social gulf between themselves and their fellow labourers, sustained by directing authoritative abuse downward, was an effective means of guarding their tenuous space. This disposition was easily adopted if we consider that the constant abuse from the *usta* produced a low self-esteem, which, coupled with an insecurity with regard to their position, manifested itself in hostility.[36]

Following Eid al-Kabir, minaret building resumed, and once again I began working with the al-Maswaris, this time on a new structure in Madinat al-Asbahi, a southern suburb of Sana'a. There were several new characters in the cast, and a few key ones from the last project missing. Majid, for instance, was completing the house project in Sanhan with the assistance of a small group of labourers; likewise, 'Abdullah as-Samawi had been put in charge of a five-day project in a tiny hamlet outside Sana'a to build a five meter high minaret in brick, and for some time afterwards had suffered poor health; and Saat stayed behind in the village with his bride in order to 'learn about the joys of marriage', or so I was told. Aside from several fresh faces among the labourers, there was a new key figure, Gayyid, who, like most of the crew, was from Dhamar. He was in his late twenties, slight of build, a bright and witty personality, and claimed that he himself was an *usta* when working back in his home region. While working for the al-Maswari, he assumed the title of *thana'a*, commonly understood in this context to mean 'the second below the *usta*'. At the beginning, he oversaw the organisation of tasks among the labourers, and was primarily responsible for filling in the walls. Interestingly, the preposition *bayn thaniyah* from the same root as *thana'a* and meaning 'inside, among, between', is an apt description of both the *thana'a*'s position in the hierarchy between the *usta* and the labourer, and of his main task of filling in between the finished wall surfaces.

With the changes to the team at the beginning of the project, I found it difficult to classify the individual builders in terms of their rank and duties.

This seeming confusion was actually an opportunity for me to break down some of the preconceived notions I held about the apprentice's position and about the division of labour in general. To begin with, there hadn't been a *thana'a* at the 'Addil minaret, and the sudden appearance of one at the al-Ihsan minaret in Madinat al-Asbahi was initially odd. The filling in of the wall space at the previous minaret had been accomplished by one of the lower-ranked *asatiyyah*, usually Majid or 'Abdullah, and not by someone with a specially designated title. What was obvious was that, with both Majid and 'Abdullah absent, Muhammad and Ahmad required senior assistance on the site and hired Gayyid through the Dhamari connection. Majid never returned to work on this project – he was *ta'ban*, tired, worn-out, and needed to rest at home. When 'Abdullah finally returned, Gayyid maintained his position, essentially fulfilling the role of the fourth *usta*, but without the title and without the due respect. Although he was an *usta* when on his home turf, he was not entitled to call himself so while employed by the Bayt al-Maswari. The prestige of his portfolio of small village projects was re-contextualised in the company of such a prominent House of Sana'ani builders. Working for them meant a decline in status in exchange for a gain in valuable experience and wider recognition.

With Saat on temporary leave setting up home, a prospect I had great difficulty visualising, the vacuum at the top of the labourer's ladder attracted fierce competition between several contenders, namely 'Muhammad the Tobacco Chewer' and young Ibrahim, son of Ahmad al-Maswari. Not showing great academic potential, Ibrahim abandoned his secondary studies after Ramadan in order to work full time on the minaret site. He quickly carved a niche for himself by publicising his new dedication to the family profession and determination to one day become an *usta* like his father. During the construction of the stone base, he worked closely alongside Ahmad, eager to make the right impression and to closely ally himself with the al-Maswari authority. His tasks mainly consisted of fetching dressed *habash* stones from the *muwaqqis* (stone mason) to be used on the outer face, and to mix the lime-based mortar for his father who set them. With passing time and rising levels of confidence, young Ibrahim became more adept at ordering his fellow labourers around than performing his own duties. Due to his youth, lack of trade experience, and unjustifiably inflated ego, he had not earned any respect from his colleagues, who mostly ignored his commands and frequently challenged him. When confronted, he retreated, cowering under the protection of his father and further discrediting himself in the eyes of others.

Several weeks into the project, Ibrahim was given the responsibility of infilling the stone walls of the base along with Gayyid. Unlike Gayyid, however, Ibrahim was not given the title of *thana'a*, but rather he was a *rassah*, a local highland term used to denote the person who fills in the walls, and, speculatively, a word derived from the verb *rassa* which means 'to join together' (perhaps joining the two wall faces with infill). Theoretically there was no attribution of authority with the task, but he became increasingly obnoxious and less conscientious about his tasks. He was reprimanded regularly by both his uncle and father, and frequently-registered protests from Gayyid and Muhammad the Tobacco Chewer led to Ibrahim's temporary banishment from the top of the structure to work below with the other 'lowly' labourers. His uncle complained that his irresponsible nephew was still lacking reason and therefore there was no sense in disciplining him. His logic echoed the common belief in Yemen that reason, *'aql*, is not the product of training, but rather it must be present in order to proceed with training.

The increasing height of the structure necessitated taking extra precautions against accidents, as a fall from as little as ten meters onto a pile of sharp stones can be unforgiving. On windy days Ibrahim often wore a narrow strip of braided black goat's hair tied around his ankle for protection. This type of *hirz al-djinn* is commonly worn around the wrist or ankle, and protects the wearer against evil and mischievous *djinn* that are popularly believed to travel on the blowing winds. The braided goats hair, known as *za'al*, must be imbued with protective properties by a person with special powers such as religious men, *sayyids*, or by a *mushawath*, who are known to be magicians or tricksters. All of the men at the building site believed that *djinn* have the power to meddle in the lives of people, but many preferred to leave their protection in the hands of God. Ibrahim was occasionally teased that *djinn* only come to 'clean' people so why should he worry? Insults of 'uncleanliness' have strong connotations of being spiritually impoverished and neglecting one's religious duties as a Muslim, namely ablutions.

Muhammad the Tobacco Chewer, 'Muhammad T.C.' for short, had been eager to prove his mettle since the 'Addil minaret project, and now, with Saat's absence and Ibrahim's misconduct, he had manoeuvred himself into directly assisting Muhammad al-Maswari (see plate 44). His responsibilities were essentially the same as those I have described for Ibrahim, but he performed with markedly more vigour. When questioned, Muhammad T.C. was surprisingly vague about what he wanted to do for the rest of his life, but he was certain about wanting many children and the free time to go hunting with the gun that he kept in his village. In the meantime, his serious motivation to impress the *asatiyyah* and his militaristic drive to control positions on the building site made him a strong competitor for a career in the trade. It is important to acknowledge that for most of these young boys, the prime motivation for acquiring this position was the associated power, not the formal acquisition of trade knowledge. This latter aspect of the apprenticeship was, in a sense, a by-product, albeit a crucial one, of gaining and maintaining that top labourer's post. Because the trade knowledge was not objectified and not discussed, mastering the hierarchies of power in the system became the most tangible goal for the initiate.

147

PLATE 44
Muhammad the Tobacco Chewer, one of the contending apprentices working for the Bayt al-Maswari.

Many things had changed by the time Saat returned from the village. The base of the al-Ihsan minaret had already been completed and the brick tower was well underway. Much to his chagrin he found that the position he had toiled for so many years to secure was no longer waiting for him. He now had to share the limelight with Ibrahim and Muhammad T.C., and was answerable to yet another person, Gayyid, who, as *thana'a*, inherited some of Saat's former responsibilities. With time, and mainly on account of some clever tactical manoeuvring by Muhammad T.C., Ibrahim lost most of his uncle's remaining trust and was increasingly forced to the periphery of the competition. Muhammad T.C. ruthlessly took advantage of every given opportunity to publicly expose Ibrahim's weaknesses and laziness, but it was Saat's quiet perseverance that won out. Within a couple of months of his return he had managed to re-appropriate his central position distributing tools and materials to the asatiyyah at the top. However, shortly after, when Gayyid fell ill with malaria, Ibrahim was once again promoted to *rassah*, meaning that Saat was obliged to serve him as a supplier. These shifting events clearly frustrated Saat, but, not being a member of the *Bayt* (House), he was powerless to contest such decisions.

In summary, the hierarchical positioning among the labourers was fluid, and there was no guarantees that one would maintain a specific posting with its affiliated responsibilities and power. This uncertainty, in combination with the verbal and psychological abuse inflicted by the senior builders, produced anxiety in the aspiring young labourers which often became manifest in the forms of hostility directed toward colleagues and in particular toward potential competitors. Thus the communication of work-related directives dominantly took the form of reprimands, rather than explanation, at all levels of the hierarchy, from top to bottom. Reprimands were curt and mostly involved name-calling such as 'dog', 'donkey', 'brainless' (*ma fiish mukh*) or '*habashi*' (Ethiopian, with derogatory reference to being 'black'), and threats that they would be sent back to their *qabali* (tribal) village. Because of the fluidity between positions below the rank of Master Builder, between their associated responsibilities, and between those players who may have held these positions at any given time, it was difficult to speak about 'an apprentice' or 'apprentices', or even about the specific traits of the 'apprenticeship'. However, in the remainder of this chapter I will demonstrate by reference to the tasks and responsibilities assigned to these young men that there is a regulated system for the transmission of trade knowledge which can rightly be referred to as an apprenticeship.

Slippery Categories: towards a theory of concepts beyond language

The absence of a distinct categorical classification for the 'apprentice' was, as I pointed out earlier, initially difficult to accept. Everyone, including members of the general public familiar with the profession, acknowledged the fact that this particular builder in question worked alongside the Master Builder and acquired both trade skills and expert knowledge. It was also recognised that participation in, and fulfilment of, this privileged stage was necessary for the passage from 'lowly' labourer to the status of *usta*. Because no specific term was popularly used to describe either the person or the system in which they were engaged, I was forced to conclude that both were recognised and conceptualised, but the concept(s) remained locally unnamed. This seemed to present a serious challenge to many of the language relativity claims put forward by Edward Sapir in *Language*, more specifically his assertions that:

> Thought may be a natural domain apart from the artificial one of speech, but speech would seem to be the only road we know that leads to it;[37]

language, as a structure, is on its inner face the mould of thought;[38]

and especially the notion that:

> the single word expresses either a simple concept or a combination of concepts so interrelated as to form a psychological unity.[39]

If Sapir were correct, how could the builders seemingly conceptualise a system that they both inhabit and reproduce, but that they did not name?

At one point I determined that there may be some relation between this and a number of other observations that I had taken note of during the course of my studies. On my numerous walking excursions in the mountains I regularly asked the local populations about the names of specific plants, birds, and insects, curious to know the locally used Arabic terms. Before long I was forced to conclude that, in general, these taxonomies were very poorly developed among the every-day language users that I had encountered. Repeatedly, plants of every kind and size, ranging from flowers to trees, were referred to by a single term, *shajar*, which translates as 'trees, shrubs or bushes'. Aside from the word for rose, *ward*, which also translates more generally as 'blossoms or flowers', the few other specific names that I managed to muster included the name for the tamarisk tree, *'athl*, and for several plants commonly used medicinally, but their names often displayed regional variations. Likewise, bird names in common use were also relatively few including *hid'a* for kite, and *ghurab* for raven. For the most part, any bird was described as *tair*, the general word for 'bird', or as *'usfuur*, a term used to refer to any small bird, but more specifically meaning 'a sparrow'.

I was slightly more optimistic about the taxonomy for insects that regularly included specific names for ants, *nimal* (sing. *namlat*), flies, singular *tateer*, butterflies, *farashah*, and such phonologically amusing names as *jundub* for grasshoppers, and *basharat al-a'nab*, harbinger of the grapes, which was a favourite catch for children in the Springtime. On one memorable outing in the countryside, however, while catching my breath next to a large, rock-hewn cistern in a mountain village, and surrounded by a small crowd of curious villagers of all ages, I openly asked for the Arabic name of the large red dragon flies that hummed over the surface of the water. There was a long pause of silence before one young man stepped forward and asserted confidently that these were "*Tuyuur*", 'birds'. A general buzz of consensus followed from the crowd confirming that this was a reasonable reply. "BIRDS!" I exclaimed with surprise, "then what are those flying above us?" pointing to the circling kites. "Those are also birds" responded a second older man in a calm, reassured tone, "these here are small ones".

My most industrious investigation into categories was the search for colour terms, which I will present briefly here to demonstrate the fluidity of categories, the importance of cultural saliency for the development of taxonomies, and how these combined might relate to the lack of a specific term for 'apprentice'. Admittedly, I had been intrigued by the famous study by Berlin and Kay on *Basic Colour Terms* (1969) prior to arriving in Yemen, but I took an active interest through a series of incidental experiences. The first of these occurred when I came to Yemen in the Spring of 1996 to pursue Arabic language studies before officially commencing fieldwork in the Autumn. After nearly three weeks of courses, four different teachers, and numerous conversations with other Yemenis that in some way involved the subject of colours, I remarked in my journal that:

> the spectrum of colour names is very limited. Terms for black, white, red, blue, yellow, green, brown, grey, and less commonly, orange, pink, and purple, seem to comprise the full extent of the average speaker's colour categories. When pressed for more specific classifications such as those we commonly find in English (for example *emerald*, *forest*, and *olive* as definers of different shades of green), I find that there is little more than *ghamiq* and *fatih* to discern dark from light shades respectively.

In related discussions, I was informed by several persons – some of whom were highly educated in a formal (Western) sense, spoke fluent English, and were at least vaguely familiar with the palette of English colour terms – that "Arabs [read Yemenis] speak more *generally* about colours than you do in the West". There was a broad consensus that an English speaker's system of qualifiers for different shades was overly complicated and not particularly valuable. The existence of a comparatively diverse vocabulary for colour terms used in spoken European languages may coincide in large part with such historically-rooted cultural pursuits as painting, and the corresponding fabrication of paints, pigments and dyes. More immediately, it reflects post-Enlightenment rationality and an obsessive occupation with scientific classification. It must be recognised that science "constitutes a rather specialised activity of thought, one that is hardly required for an apprehension of humankind's immensely rich and varied everyday world".[40]

Throughout my field studies, when presented with the right circumstances I would indulge in the, perhaps ethnocentric, exercise of asking male friends, my co-workers at the building site, and children of various ages to describe colours for me. Briefly, what I determined was that the boundaries between general colour categories were often blurred, and terms were casually exchanged for others. Also, when asking someone to name the colour of an object which was presented before us both, there was often a pause of several seconds before a response was issued, and this was occasionally modified by a second reply. A 'blue' which I, as an English

151

speaker, would categorise as being "definitively blue" might be called either 'blue', 'green' or 'brown'; for example, even a cloudless blue sky, properly called *samawi* (meaning sky blue, and one of the rare terms used to qualify a shade), might be referred to by different speakers as 'brown'. Likewise, green may also be identified as 'blue', and less often as 'brown'; and red was referred to as 'red' or 'brown'. Despite the frequency with which the term 'brown' was used, and given the large vocabulary in Classical Arabic for describing different shades of this colour in particular, its expression was limited to either the standard term, *bunni* (coffee-coloured or brown, from *bun* meaning coffee), and qualified as either being dark or light, and, infrequently, *dam al-ghazelle* (gazelle's blood) to describe reddish-browns such as sepia tones.

The not inconsiderable data which I recorded on the use of colour terms amongst predominantly Sana'ani males led me to deduce a series of continua existing between the popular use of the terms for red and brown, between brown and blue, and between blue and green – and occasionally a blurring between the terms for brown and green as well. It must be noted that my contacts were almost exclusively limited to the male population, and I have been advised by female colleagues who were in regular contact with the local female population that the Sana'ani women use a larger vocabulary of colour terms than the men, more often employing the terms for orange, pink, and purple. Whether the women were equally disposed to blurring these categories was left undetermined. Nevertheless, the question remained as to whether the existence of continua between colour categories, prevalent amongst those I spoke with, indicated that the average Yemeni male's 'cognitive spaces' for colours were equally fluid.

For instance, if one English speaker in conversation with another uses the term 'blue' to describe the colouring of a subject, he/she can be reasonably confident that the listener has conceptualised a 'blue' subject, and not a green or a brown one. The term 'turquoise' is more ambiguous as to whether it refers to a subject which is more blue than green, or vise versa. The same English speaker can attempt to pick-out a more specific cognitive space in his listener by qualifying the blue as cobalt, indigo, periwinkle, etc., depending on the levels of sophistication of either the speaker's or listener's colour ontologies. In Sana'a, however, when using the term *azraq*, blue, to describe the colour of an object in conversation it remained uncertain in my mind as to whether the listener was also 'thinking blue', and not alternatively conceptualising green or brown. In effect, the issue is whether the shifting lexical categories indicate equally undefined conceptual classification of colours.

Daniel Dennett makes the seemingly reasonable assertion that a concept "is an internal label which may or may not include among its associations the auditory and articulatory features of a word (public or private)", but goes on to suggest, more problematically, that "the first concepts one can manipulate ... are 'voiced' concepts, and only concepts that can be *manipulated* can become *objects of scrutiny* for us".[41] How then does this latter assertion accommodate certain practices exhibited by Yemeni men who have not acquired an extensive vocabulary of colour terms but scrutinise between different fabrics in the tailor's shop according to distinct preferences for certain colours, and even specific shades. Choice was exclusively made physically through gesture and pointing, and not by specifying any colour term verbally.[42]

Clearly the importance of determining correlations between cognition and language was pertinent only to me at the time, and of no consequence to my Yemeni colleagues and contacts. My initial intrigue with comparing ontologies admittedly stemmed from my own implicit assumptions about the world and how these become represented in our language categories. This positivist, and possibly innate,[43] form of assumption-making that continues to burden much anthropological inquiry might be referred to as a form of 'psychological essentialism', consisting of the belief that "the world has a natural order that is independent of the observer", and "the categories – and words referring to categories, such as common nouns – map onto that structure".[44] As pointed out by John Lucy, comparative colour studies, such as that by Berlin and Kay:

> still retained the lexical orientation of the earlier era's fundamentally Western conceptions of 'colour' characteristics. Rather than working from a comparatively induced typology of patterns of language-world relationships, it showed instead the distribution of languages relative to a fixed set of parameters drawn from the western European scientific tradition. So despite its comparative orientation, it actually washed out linguistic differences and suggested that languages merely 'reflect' or 'map' reality.[45]

As confirmed earlier, colours did not constitute a culturally salient feature in conversations among (the average population of) men in Sana'a, and this was reflected in the comparatively weak development of the taxonomy of basic colour terms employed, as well by the fluidity between classificatory boundaries and concepts. For example, *akhdar* (green), *azraq* (blue) and *bunni* (brown) were frequently used in an interchangeable manner in order to describe any one of those colours, which, by comparison, would be differentiated by average North American or British speakers. Colours need not be fixed by the language used by most men in Sana'a because they need not be specifically fixed as distict concepts in the thought of the average (male) speaker. Seemingly, the cultural context makes no requirement of this.

Interestingly, and by contrast, I recorded the names of eighteen different types of bread consumed in Sana'a which were regularly employed by the average local speaker. In English, they were all 'bread' to me, but as a learner of Arabic it was imperative that I know the names of those that I bought regularly. My confusion of terms elicited correction, and occasionally ridicule. Conceptually, for the average Sana'ani, these were not generically lumped together as *khubz*, 'bread', but each was distinguished by its particular quality and denoted by a distinctive term. As Maclagan notes, "a large vocabulary distinguishes dozens of different types according to the grains used (wheat, barley, sorghum, millet, maize, lentils and combinations); whether leavened with yeast, fermented, or not; and the method of cooking – as a pancake on a griddle, spread by hand on the inside of a cylindrical oven, or preformed and applied there on a cushion".[46] The following list of 'bread' types available in Sana'a which I compiled is by no means a comprehensive inventory:

jaheen	corn bread
qafuu'a/bilsanah	lentil bread
tameez	type of unleavened bread
khubz bur	wheat bread
maluuj	large circular flat bread (eaten with *saltah*)
rashuush	large circular flat flaky bread, often sprinkled with seeds
kidam	small round heavy bread, introduced by the Turks
kidam ahmar	brown *kidam* bread
ratab	soft oily bread
sha'ayar	barley bread
lahuuh	like Ethiopian *injera* bread
khaas	circular white flat bread
ruwtee	white French sticks of bread
khubz dukhin	pearl millet bread
maluwah	large thin flaky round bread (eaten with fish)
thamuul	wheat bread made with egg (eaten during Ramadan)
zalabeeya	bread fried in oil (eaten with *sahaawiq*)
khubz taawah	thin bread with oil, fried like pancakes (eaten with tea)

Recognition of culturally distinct systems of classification, like the example above, lends support to Lucy's position advocating a reconsideration of the Whorfian relativity hypothesis, and "beginning with the language structure to ask what it suggests about the implicit construal of reality" in a given population.[47] However, in order to understand how an entity such as what I have called 'apprenticeship' can exist, be recognised, aspired to, and reproduced through a distinct process enacted by agents, without itself being a 'voiced concept' in the language structure, it will be necessary to expand this approach and challenge the prevalent notion of language being the principal medium through which concepts are made manifest.

Firstly, adopting an hypothesis defended by Atran, Sperber, Cosmides & Tooby, and Boyer,[48] I likewise maintain that the human capacity for the construel of folk taxonomies, and not the taxonomies themselves, is the result of cognitively constrained capacities operating *a priori*, and independently of, the cultural context. These 'modularity-of-mind' theorists refer to this as a domain-specific cognitive capacity. It is the innate propensity to categorise the world into distinct, and interrelated, concepts that provide the means for the development of theories, cultural representations, and such 'higher' forms of scientific classification, and not the reverse.[49] Gelman, Coley & Gottfried have termed these concepts 'essentialist assumptions', as described above, and claim that an essential bias is carried from childhood throughout adult life. The most abundant evidence supporting an essentialist world-view is in the categorisation of living kinds.[50] In earlier work, Boyer extended the idea of an innate human capacity to conceptually construct essentialist assumptions for living-kinds to naturalised positions of agents in traditional contexts such as religious ones. He write that:

> positions are conceived in terms of essence, in much the same way as living kinds. The accession to a certain position is therefore conceived as the unfolding of underlying properties rather than as a contingent change in people's activities, even when such a change is the only means to recognise the position. The representation of positions as natural makes it possible to conceive certain causal links as affecting certain persons only.[51]

The role and position of the top labourer (read apprentice) working alongside the *usta* in Sana'a was popularly conceived primarily as 'process' and not as part of a rigid morphology of rank. Therefore, *thinking* about the 'apprentice' was *thinking* (about) the 'process' of this particular system. Because this 'thinking' was clearly not encoded by specific lexical terms, it was therefore not guided, at least not entirely, by a propositional form of knowledge. Evidently, if the position and role were 'recognised, aspired to, and reproduced by agents', then it was not only thought about, but importantly it was a concept that was also scrutinised and manipulated at some level. To 'recognise, aspire to, and reproduce', requires intentionality, or an 'intended-towards', which equates to a conscious awareness of the concept or thought. This would seem to contradict Jackendoff's argument that thinking, unaccompanied by lexical terms, and more particularly phonetic form, is not available to a conscious level of thought.[52] It is therefore necessary to go beyond language in investigations of conceptualisation, and to recognise that there may be distinct kinds of conscious thought that are largely independent of language, and that "at most the thesis of linguistic determinism would be a thesis about the relation of language to just that part of the cognitive system specialised to propositional representations".[53]

Evidently different modes of training produce expertise in different domains of knowledge, some of which are not expressed through language. Recognition of this mandates "a dynamic approach to word meaning and thought more generally, involving interactive modular components".[54] In studying the skilled performance of an American knife maker in relation to his verbal communication with his client, Keller & Keller concluded that "words are functioning at the interface between cognitive systems and at the interface between minds".[55] They observed that "once production of the blade has begun a dynamic interaction between imagery and material developments guides the production process".[56] The skilled concepts of the knife maker, like that of the builder, are vested in visual and sensorimotor domains of knowledge. Verbal directives interface with, and can serve to modify, the concepts of these domains, but are not the source of their formation.[57] 'Making', not language, is the source of their knowledge.

To summarise, an inquiry limited to registering and analysing propositional statements would be severely constrained as the traditional builders were not proficient at verbally explaining *how* they know.[58] This limitation instigates a modular approach to *thinking about thinking*.[59] Establishing an analytic framework which considers not only propositional concepts in language, but sensorimotor representations in performance, imagistic representations in the 'mind's eye', and possibly others, better situates an investigation of training-learning processes.[60] In the following sections of this chapter I will explain 'modularity' and develop this theory in order to demonstrate *how* the building apprentice acquires skills, dispositions, and world representations through the existing training process, and importantly, how he learns to 'make'.

A Comparative Study of Training-Learning: Sanaʿa and Quebec

I now return to the account of my first day at the building site, sitting uncomfortably on top of the wall next to Ahmad al-Maswari, waiting nervously to be assigned a role. After digesting my presence for a few minutes and coming to the realisation that I was indeed serious about building as a labourer on his team, he set me to work. For the remainder of the early morning I worked at the top with Saat and another labourer, Adil, stacking bricks in rows that lined the curve of the stairwell wall, conveniently placed so that Saat could pass them along to the *usta* when the pace of work quickened in the late morning. Not once was my task or role explained to me. Ahmad simply gestured with a forward extension of his right arm and an opened hand toward the work in progress, accompanied by a barely audible command, "*Yallah, ishtaghal maʾahum*", "Go on, work with them". Initially my mind raced, scrambling to make sense of my new environment, the people, the task. Intuitively I had to negotiate my physical relation *vis à vis* my two work-mates, and determine how to effectively and economically co-ordinate my body's performance in passing, hauling, and stacking the bricks.

Throughout my work periods on the school and the minarets I made a concerted effort to practice a hermeneutic phenomenology, simultaneously immersed in the practices of labouring and the practices of reflecting upon the essence and mechanics of the many processes I was directly experiencing through either my engagements or observations.[61] I kept mental and written notes, as well as 'embodied' notes (and not only aches and bruises), on these as I worked. My on-the-spot learning to cope with, and perform my very first task depended on my ability to tap a vast number of different information sources. These included, in a non-hierarchical order: a recollection of being told at an early age by my father "don't pick things up with your back. Use your legs. Bend them like this ... " as he proceeded to demonstrate; a conjuring of the imagery and feel of having worked closely with contractors and carpenters on the construction sites of my own built designs, which served to gauge my expectations of the temperament of these builders, and anticipate some of the occupational hazards on a building site; a practised capacity for mimicry from my sporting and gymnastic experience which enabled me to observe and subsequently calculate an approximately accurate performance; a reliance upon learned and practised spatial-temporal co-ordination in order to judge distances and the tempo of the co-operative tasks; an intuitive recollection of the general laws of physics guiding my judgements of what was and wasn't physically possible, for example, in arranging the high, and precariously stacked, rows of bricks at the top of the stairwell; and a recollection of my recent visits with these same builders, which informed my early anticipations about the relations I would have with them, and about the overall nature (both physical and technical) of the work. When I took Saat's place and straddled above the staircase with one foot on the wall and the other on the central column, I could also recall my parents nagging me as a young boy to "get away from the edge" of those garden walls and high places I loved so dearly to walk along.

Within a framework evoked by these varied concepts, I formed an initial working drawing (or mental map) of the context in which I was both present and participating. Although I have described the components of this frame with written words, very few of the actual concepts were propositional in the sense that they were *not thought about through language*. Aside from the cited precautions issued by my parents and piers, the bulk of the concepts that I relied upon to situate myself and co-ordinate my actions were primarily sensorimotor or image related. When studying how my fellow builders likewise situated themselves and co-ordinated their actions, I had access to their concepts either through what they told me or through what I could observe performatively or emotionally, but by nature

157

of our cognitive limitations as humans, I was denied access to their imagery-based conceptualisations. To get some notion of this, I had to assume that I was able to conceptually share some of their everyday imagery by virtue of living in Sanaʿa, by regularly engaging with them in work and social relations, by learning about the country's historical, political, and cultural context, as well as by familiarising myself with the life histories and present circumstances of the individual builders with which I could experience some (perhaps limited) level of empathy. Therefore my ensuing representation of the builder's teaching-learning process is predominantly an account of my recorded observations of my fellow workers, overlaid with my own, mainly image and sensorimotor-based, concepts that guided me through my own training.

A large component of the apprentice's training was merely a continuation of the disciplinary formation of a labourer as described in the last chapter. As already stated repeatedly, expectations came without explanation, and failure to produce a satisfactory performance solicited harsh reprimands of curses and name calling, "*Ya kalb*!", "*Ya Yahud*!" (Dog!, Jew!). Due to the physical proximity of his work station to that of the Master Builders, and the increased responsibility for co-ordinating the tasks and the tempo of his colleagues' production, the apprentice bore the full brunt of the *usta*'s anger and frustration, and seldom enjoyed any praise. Courtesies were essentially non-existent in the builders' spoken discourse – never an accompanying 'please' or 'thank you' – and the abrupt, and sometimes aggressive, intonation of requests transformed them into commands. Illocution was far weightier than the locution. In the later afternoon, once the *qat* had taken effect and the builders were completely absorbed in their individual tasks and thoughts, sometimes pausing to drag on a cigarette and stare vacantly with wide glassy eyes toward Jabal Nuqum in the distance, the vocal commands regressed to barely intelligible grunts.

The apprentice's work station at the top of the minaret staircase was narrow and cramped. Stacking the bricks created an enormous amount of throat-irritating dust, and the worker in this position was under frequent assault from projectiles of brick chips and globs of mortar sent flying by the chisels of the seemingly oblivious *asatiyyah*. Casual visitors to the site and stockpiles of materials invaded his already tight space and interrupted the progress of his tasks. He was primarily responsible for preparing and distributing tools and materials to the builders, and tempering the flow of required materials passed from the ground and up along the staircase by the labourers. He was obliged to arrange the arriving supplies of materials along the top of the staircase, sometimes with the assistance of another labourer. He soaked the bricks in a water tank prior to distribution, and mixed buckets of concentrated lime mortar when required. The apprentice had to anticipate the imminent demands for materials by the *asatiyyah* or await their commands which he himself echoed down the stair shaft, hollering for the labourers to pass along more *"Yajurr!"*, *"Khalta!"*, *"Asabi'!"* (Bricks!, Mortar!, Branches!). His own abruptness and harsh intonation came to mirror those of his Masters.

In contrast to the apprentices I worked with at the building site for the school who were granted relatively greater autonomy and were more regularly involved with aspects of *making*, the role of the apprentice working for the Bayt al-Maswari was primarily to serve the Masters and facilitate their tasks. Occasionally there were windows of opportunity for this labourer to become more directly involved in the actual building process, normally when one of the *asatiyyah* was absent or when the construction reached certain stages that required the execution of mundane tasks preferably avoided by those of higher rank. These supplementary responsibilities included the polishing of the exterior brick face, completing the infill between wall surfaces, and most interestingly, the carving of bricks to be used in the decorative relief geometries. Carving bricks is one of the earliest skills that any ambitious labourer must master in order to be delegated subsequent tasks. The brick is the smallest basic unit in the composition of the structure, and metaphorically constitutes the basic unit in the accumulation of knowledge related to *making*.

The nature of my own lessons in brick carving paralleled those I had observed for other labourers both at the minaret and at a number of other building sites. Not surprisingly, the transfer of this knowledge was effected through the performance of the skill and accompanied by minimal verbal directives. Indoctrination was predominantly by process of observation and mimicry. The first standard brick shape that I learned to carve was popularly called the "Madame", in a rather vulgar reference to the deep V-shaped incision that is made along one of the square faces. It is one of the more complex shapes to manufacture, requiring four angular cuts that result in an overall V-shape of the principal face. When the brick is laid on its back with the incision facing upward, it is also called *bud'* which translates to 'vulva', and comes from the verb *bada'a*, meaning to 'cut, slash, or slit open'. Alternatively, when the brick is set with the incision facing downward, it is called a *makhta*, which I suspect may be derived from the verb *khata* meaning to 'step, pace, or walk', befitting its appearance.

Majid, the son of Muhammad al-Maswari and third ranking *usta* on the site, was my teacher for this task. He motioned to me to squat down facing him while he chose a brick from the stack at his side. He held it in an upright position on the ground between us with his left hand and gripped the handle of his adze in the other. The only spoken explanation that he offered before proceeding was that the first cut "must begin at the corner of the brick", while simultaneously pointing to the spot with the blade of his tool which he positioned in such a way as to evoke the desirable angle of the cut. There was no use of quantitative values to describe the geometries of the required angles. Majid looked me directly in the eyes to be sure that my full attention was on him, then raised the adze above his shoulder and came down upon the brick with an exacting blow that eliminated the corner. This was immediately followed by four or five rhythmic strokes of the blade to even the cut. With his eyes focused on his task, he rotated the brick ninety degrees and again, holding it firmly, he reproduced the same carving action on the far adjacent corner. He held the brick in the air in front of his face, level with his eyes, to quickly verify the angles, then placed it back on the ground slightly closer to him and proceeded to make the central V-shaped incision. This required two further angled cuts whose slopes mirrored those he had produced on the outer edges. The strokes of his adze were decidedly smaller and less forceful here, mindful not to make the incision too wide, which he satirically called a "French Madame", a remark steeped in allusions to the foreign 'blue' films he had been watching on his satellite television.

Satisfied, Majid handed over his completed model for me to study, then passed me a second adze and a new brick that he had selected from the pile. I was slightly nervous during my first attempt, and wanted to make a good impression in fear that I might not otherwise be considered for subsequent tasks of *making*. I held the brick as he had demonstrated and carefully placed the blade of the adze on the corner to visually set-up the forty-five degree cut in my mind. I drew back my arm keeping my eye on the imaginary cut-line where the blade had been, then issued a blow followed by a staccato delivery of several others. Majid silently watched. I turned the brick and repeated the cut on the next corner, held it up scrutinisingly, then proceeded to make the incision. My final product was examined by my teacher, and without speaking, he directed my attention back to the brick and slightly modified my cuts with his own adze. Again he handed it back to me so that I could study his amendments, before giving me another brick for trial number two. I shaped my next "Madame" more confidently, but was told that my incisions were too wide with playful references to a host of "Mesdames" of various nationalities. My third attempt clearly pleased Majid, and he held it up for the others to see, proclaiming with a broad grin that I "couldn't be Canadian, but must be Yemeni!" (see plate 45).

For the remainder of that morning I worked at the top of the walls carving bricks which were being used in the decorative relief patterns. I was given further instruction on how to select the right bricks for carving, being told that the more homogeneous red ones were best suited for the task and that I should reject those that had been over-baked or that had black content, as these were generally hard and brittle. For now, I was restricted to producing this one shape, and would be taught how to manufacture others later. Each shape had a distinctive name, and these were universally used and understood by the builders that I spoke with in Sanaʿa. Some builders occasionally employed their own home terminology, but the following comprises the standard set of brick types (see figure 8):

PLATE 45
'Abdullah as-Samawi carving
bricks.

bud^c	vulva, slit; right side up and V-shaped.
makhta	the walker (?); upside down and V-shaped.
maqruun	linked or united; completing the V-shape of either the *makhta* or *bud^c*.
a'sar	left-handed; referring to its position in a relief composition, extending a leg of either the *makhta* or *bud^c*. Also pronounced *aysar*.
ayman	right-handed; referring to its position in a relief composition, extending a leg of either the *makhta* or *bud^c*.
takbuush	likely from *kabsh*, meaning 'prop or support'; set decoratively in a band one next to the other. This term is also used to describe the bricks cut at an angle to form the corners of an octagonal plan.

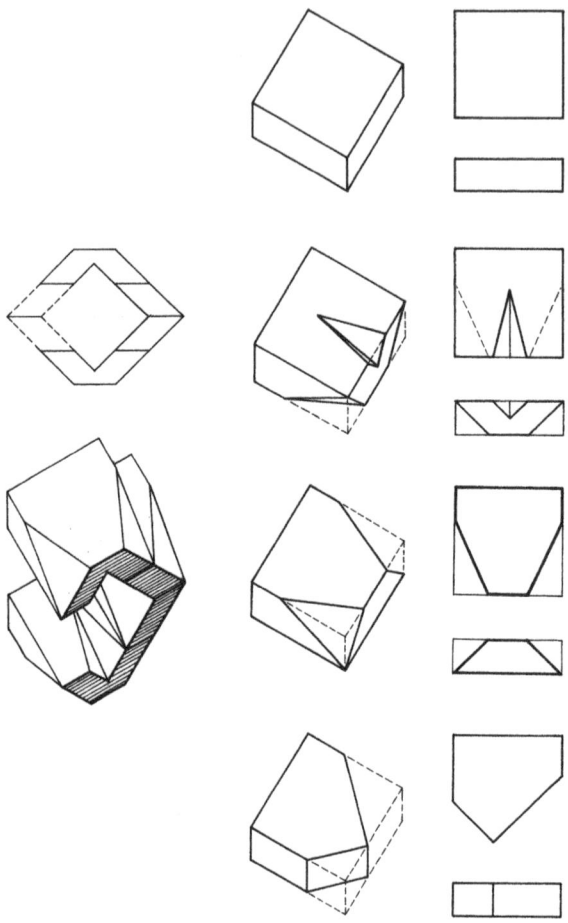

FIGURE 8 (Right) An un-cut *ajurr* brick and three brick cuts (from top to bottom): a) *bud'*, or "madame"; b) also *bud'*; c) *takbuush*. (Left) Diamond-shaped decorative assembly (or *diblah*) of cut bricks.

In contrast to the individual names for the brick types used in the execution of the decorative relief patterns, the patterns themselves did not have specific names which were collectively shared by the Sana'ani building community. The al-Maswaris referred to all of their brickwork designs by the term *musalsal*, which means 'chained or continuous', and which aptly describes the common theme of interlocking squares, diamonds, and chevrons in the various patterns.

Although not strictly part of the training of a traditional builder, some young labourers also assisted the *muwaqqis*, or stone mason, in dressing the machine-cut stones as part of the learning process. In his study on traditional architecture in Sana'a, Al-Sabahi suggests that being a stone mason is one of the five stages in becoming a Master Builder.[62] However, my findings with the Bayt al-Maswari were that the *muwaqqis* was an independent craftsman who was hired contractually and paid a daily wage, *al-mushaqah* (see plate 46). A labourer who was intensely trained in this skill would likely remain in that particular trade, as the *muwaqqis* was one of the better paid members of the work team and had considerable autonomy.[63] One young labourer who did periodically assist Yahya Faraj, the al-Maswaris' brawny stone mason, provided an interesting case study of training-learning processes.

In assisting Yahya to dress the black *habash* stone that was being used to face the square minaret base, the labourer was meant to bush-hammer an even pattern of diagonal lines across the stone face. The bush-hammered patterning in stone is locally called *raqamah*, being derived from *raqama*, meaning 'to brand, imprint, or to stripe (as in cloth)'. Yahya did not explain to the young man how he was supposed to perform this task, instead leaving him to observe and mimic in the same manner that I was shown to carve bricks. The stone mason hammered into the face of the stone with great force and a controlled, machine-like precision. He moved the blade of his tool methodically from left to right, and from top to bottom, along the stone plane producing dense rows of diagonal incision lines. Each blow was accompanied by a slight outward thrust when the steel edge hit the stone, producing wide, deep, even-length incisions.

The labourer had watched Yahya, but had paid greater attention to the predominantly 'diagonal' overall affect of the final pattern than to the actual performance of the exercise. He then proceeded to work his own stone, also producing diagonal lines, but in rows diagonal to the stone's edge, beginning in the top left-hand corner and moving diagonally down the face to the bottom right. He continued to cover the entire plane in this manner without being corrected by the *muwaqqis*. The result was markedly inferior and yielded a much less compact pattern. The *muwaqqis*, who had been dressing another stone, examined the young labourer's finished work and commented that the incisions should be more closely spaced and that more physical force was required to produce stronger lines. He then turned to his own stone and proceeded to demonstrate once again, still moving evenly from left to right in a rectilinear direction along the face. The labourer set about the task a second time, exerting greater effort to make bolder lines, but without performing the necessary outward thrust with the steel blade, and persisting, unaware, in hammering in diagonal rows. In short, his practice was chiefly steered by his observations and assessment of a salient feature in the pattern, namely the 'diagonal' aspect. This image-based concept over-rode his sensorimotor concepts derived from watching the *muwaqqis*, and there was no use of propositional information in terms of instructions or reprimands to rectify his performance.

PLATE 46
Yahya Faraj, the stone mason,
or *muwaqqis*.

The general lack of propositional content in the formation of skilled task concepts in the Sanaʿani traditional builder contrasted sharply with the teaching-learning processes of a professional carpenter in Quebec. The Montreal carpenter whom I interviewed, Eric, entered the trade for his "love of wood, and especially the smell of fresh-cut timber", and because he derived a great satisfaction from working with his hands. He told me that he regularly reads books and magazines related to woodworking for inspiration in his designs and to gain new knowledge about craft techniques. In order to obtain the trade cards that enable him to work professionally as a carpenter in his home province, Eric enrolled in a specialised trade school. The curriculum combined both theory and practice, with classroom sessions in the morning focusing on plan reading, technical drawing, site calculations, surveying, and the operation of tools; and practical sessions in the afternoon designed to exercise what had been learned in the morning. The course content began with instruction on how to use basic hand tools for crafting simple objects, and progressed to the use of more elaborate machine tools and the construction of a staircase, considered the most complex component in erecting a timber-frame dwelling. By the end of the course each student had effectively built every aspect of a house, starting from the simplest features and advancing to the most difficult.

In the classroom, the instructor first reviewed formula examples of how to approach a specific task, and then proceeded to discuss the relevant advantages and disadvantages, and elaborated on his own personal experience, thereby provoking the students to think critically about the knowledge being conveyed to them. They were encouraged to work with drawings and notes while building in the afternoon workshops, and as Eric told me, many professional carpenters continue to rely on formulas and calculations, especially when assembling staircases. Safety and medical first aid knowledge was stressed throughout as an integral part of the education. At the end of the course the students were examined on math, theory, and plan reading, and were graded on their practical work. Those with passing grades received a Dègrée D'Études Professionel (DEP), and were then required to seek a prospective employer who would guarantee them one hundred and fifty hours of work in writing. The young carpenters were furnished with apprentice cards by the Commission de la Construction du Quèbec (CCQ) once they could produce this written guarantee in conjunction with their degree.

As an officially recognised apprentice, each carpenter was required to complete six thousand working hours for certified employers in order to become eligible to sit the professional exam and receive their own trade cards. The first one thousand three hundred and fifty hours were accredited to the time spent in the trade school which is recognised to be a part of the apprenticeship. The total was divided into three equal blocks of two thousand hours each, and the graduation from one block to another conferred a raise in hourly wage rather than a structured change in the programme of skills being learned. Upon completion, the three-hour professional examination covered practical math, formulas, building theory and techniques, and plan reading (but not drawing). There was no practical component in the exam as it was assumed that all necessary skills had been mastered during the apprenticeship. Eric complained that "it was easier to find work as an apprentice than now as a professional carpenter because essentially an apprentice is capable of performing all of the necessary tasks but comes at a fraction of the price". Wages and recorded hours are centrally controlled by the Commission which also administers the final examination process and regulates the numbers of certified carpenters working in Quebec.

During his own apprenticeship Eric was under the tutelage of three senior carpenters and he remarked that in addition to learning new skills by watching and mimicking the performance of his mentors, all three craftsmen also frequently used written mathematical calculations to explain the exercises. The proportion of either verbal or demonstrative instruction varied according to the nature of the problem. Each of the three had their own characteristic method of teaching: the first carpenter spent considerable time verbally explaining the task, making thumbnail sketches, and writing out calculations for the apprentice's sake before demonstrating practically, and subsequently allowing him to make his own attempt; the second spoke very seldom, rather relying on his demonstrations to convey the information, and placing more stress on the exercise of observing and mimicking; the third and oldest of the three consistently offered a verbal explanation of the task at hand, then would invite Eric to comment on what was discussed. He would then proceed to demonstrate and have his pupil repeat. When Eric made a mistake, all three would prompt him through questioning to critically assess the apparent problem and propose a solution. If his proposal was still thought to be faulty the senior carpenters would suggest amendments to his concepts, often by relating their own personal experiences. If, however, Eric made an error with regard to something that he was thought to have been already be practised in, he was reprimanded in a sharp tone, "Tu doit savoir mieux que ça, Eric!", "You should know better than that, Eric!".

To summarise, there are clearly similarities and striking differences between the training processes that have been presented in this section: that of the Sanaʿani traditional builder and that of the Quebec carpenter. The two courses of training inculcated a *learning of making* that began with small components, and both slowly progressed to larger, more complex building assemblies. However, in contrast to the minaret builder, the Quebec carpenter was also learning to *make* drawings. Aside from the obvious fact that drawings serves to communicate the 'makers' intentions to others, the skill also provided the apprentice with a formal means to establish a dialogue between his ideas, and to objectify them during the process. Likewise, his trained ability to read drawings provided a channel for the objectified ideas of others to interface with his existing conceptualisations. In the next section, I will further discuss the role of drawing in the process of refining notions of 'spatial relations' at larger scales of *making*, with reference to architects.

The more fundamental variances between the teaching-learning processes of trades in Sanaʿa and in Quebec arose from the differences in the broader social and political contexts of the respective societies which affect popular attitudes toward education. Craftsmen in both places trained under the tutelage of a Master, but the Quebec model was far more institutionalised, with a controlled set of laws and procedures that are centrally governed, and a morphologically fixed hierarchy of categorisations. The Sanaʿani apprentice was simply a *shaqi*, labourer, without the security of a fixed position or title, and the success of his career was strongly determined by the will of the *usta*. The nature of the expert discourse, access to knowledge, and the power to question will be more specifically addressed in the following chapter. What is important to note for the purposes of this discussion, and the ensuing discussion on *making*, is the manner in which skill-related information was presented and conveyed in each context.

For both groups, the appropriation of trade skills and expertise was achieved predominantly through observing and mimicking. This included observation of varying proportions of either the physical performance of the mentor or the object produced as final product; and the mimicry involved in repeating exercises (or practices) in order to achieve the most effective and economic (i.e. skilled) performance. Evidently, propositional-type concepts about trade knowledge were considerably more salient among the Quebec carpenters, who, from their professional training, were required to incorporate an understanding of theory (as expressed propositionally in formulas, drawings and language) with their practice. This expert knowledge became manifest in propositional statements and drawings which could be publicly discussed, critically assessed, and possibly refuted or modified by subsequent propositional information. The Sana'ani builder's trade knowledge, on the other hand, was almost entirely vested in the performative processes of the *making*. For this reason the *usta* found it tremendously difficult to verbally 'explain' *what* he knows or, more importantly, *how* he knows. This limitation was largely due to the highly restricted role of propositional information exchanged during the training, both in terms of instruction and questioning, and thus in the subsequent development of trade knowledge. It not only presented a major obstacle to inquiry by an outside agent, but also limited the degree of interfacing that drawings or words might have with the builder's own established imagery-based and sensorimotor concepts of *making*. For this reason, the traditional trade knowledge was much less vulnerable to change than a form of knowledge that was objectified publicly through language or drawing and is, by extension, open to scrutiny.[64]

In the next and final section of this chapter I will focus the discussion on the chartered development of *learning to make* in the Sana'ani builder; a process that began by instilling a mastery over the 'object', and advanced the builder's abilities to manipulate the more abstract concepts of 'spatial relations' that existed between two or more components in the assembly. I will compare this training programme with that of the architect's, and will argue that such an analysis, building upon existing cognitive theory, is significant for advancing our understanding of the mind, and in particular, the workings of human spatial cognition.

Mastering Making and Mastering Space

Space is in essence that for which room has been made, that which is
let into its bounds. That for which room is made is always granted
and hence is joined, that is gathered, by virtue of location, that is by
such a thing as the bridge. Accordingly, spaces receive their essential
being from locations and not from 'space'

Martin Heidegger[65]

For cognitive linguists following Chomsky, language, like thought, is believed to have an innate grammatical structure, and they maintain that analysis of language can provide an important key to a more general understanding of cognition and to the nature of conceptual representations.[66] In this section I aim to show that this currently language-biased window can be expanded by demonstrating how a careful consideration of the skill-related performances inculcated during the course of a training programme reveals and distinguishes different types of cognitive operations involved in the production of conceptual representations, in particular those associated with object descriptions and spatial relations. In order to do so I will compare and contrast the training programme of the traditional builders with whom I worked with that of architects. Firstly, it will be necessary to elaborate upon some of the background theory before engaging with my fieldwork experience and analaysis as a building labourer in Sana'a and my own training as an architect.

The Theoretical Background

In 1983 Jerry Fodor published what has since become a very influential theory in the field of cognitive studies. *The Modularity of Mind* developed a strong functionalist position on cognitive structure considerably influenced by the work of Franz Joseph Gall (1758–1828), the founding father of the outdated and extremely controversial field of phrenology. Fodor contends that much of Gall's more valuable insight into the operations of the brain have been unduly neglected in twentieth century attacks on phrenology. What Fodor has found useful for the development of his own ideas is Gall's hypothesis of vertical faculties of the mind which he eclectically combines with other theories of faculty psychology including horizontal faculties and a Chomskian form of Neocartesianism. Fodor writes that "Descartes' doctrine of innate ideas is with us again and is (especially under Chomsky's tutelage) explicitly construed as a theory about how the mind is (initially, intrinsically, genetically) structured into psychological faculties or 'organs'".[67]

A Neocartesian psychology maintains that mental structure is mirrored in the organisation of the propositional content of mental states, and that

168

analysis of this organisation will explicate the innate mental structure. For Chomsky the model for human mental structure is necessarily the "implicational structure of systems of semantically connected propositions".[68] This theoretical framework evinces postulations of innate beliefs, or a nativism of propositional attitudes. Fodor finds this problematic and maintains that it is important for us to differentiate between this position of faculties-cum-belief-structures and the more functional position of faculties-cum-psychological-mechanisms, but the function of the faculties themselves may be innately specified in some way.[69]

The advocating of a functionalist theory of mind in philosophical sources dates back to Plato (in *Theaetetus*), although more contemporary versions propagated by faculty psychologists regard mental faculties as principally functional as opposed to being spatialised. Faculty psychology can be divided into roughly two distinct approaches: horizontal faculties of the mind and vertical faculties. The theory of horizontal faculties contends that "a horizontal faculty [such as memory, imagination, judgement, etc.] is a functionally distinguishable cognitive system whose operations cross content domains ... [and] the typical function of a cognitive mechanism is the transformation of mental representations".[70] The horizontal faculties are the implicit *architecture* of the mind and are not themselves content specific. Therefore, the judgement of the aesthetic quality of a painting and the judgement used to discern the difference between two persons' faces is derived from the same cognitive system, or faculty, in this case of judgement. Fodor believes that we must both distinguish the horizontal faculties with their characteristic patterns of transformation of mental representations, and individuate content domains in order to determine whether the operations of faculties do indeed cross content domains.

Franz Gall disregarded the prevalently accepted notion of horizontal faculties, contending that there is "no such thing as judgement, no such thing as attention, no such thing as volition, no such thing as memory; in fact there are no horizontal faculties at all ... [because] if there is only *one* faculty of (say) memory, then if somebody is good at remembering *any* sort of thing, he ought to be good at remembering *every* sort of thing". Instead Gall advocated the idea that there is a bundle of "propensities, dispositions, qualities, aptitudes, and fundamental powers" that are distinguished by their domain specificity, for example, music or mathematics.[71] Fodor has termed this notion of specific, possibly localise-able, intellectual aptitudes 'vertical faculties'. What is useful in Gall's ideas for guiding a theory of cognitive modularity is not his renunciation of horizontal faculties, which as Fodor argues most probably exist in combination with vertical ones, or even the domain types which he denoted as being vertical faculties, but rather the inherent qualities of vertical faculties. Vertical faculties are attributed with being domain specific, genetically determined, and associated with distinct neural structures. Fodor deduces a fourth quality, namely that they are computationally autonomous, which means that vertical faculties "do not compete for such horizontal resources as memory, attention, intelligence, judgement".[72]

Following his review of 'associationism', which I will not elaborate here, Fodor concludes that if a cognitive faculty is believed to have a distinct neural structure (like vertical faculties), then we must reject the idea advocated by associationists that faculties are *assembled* from an inventory of more basic sub-processes. He maintains convincingly that efforts of computational associationists to reconcile faculty psychology with Empiricism (by devising a theory of learning that would undermine nativist axioms of mental structure) have largely failed.[73] With the development of this fifth premise, Fodor then postulates his well known theory of modular cognitive systems, declaring that they must be "domain specific, innately specified, hardwired, autonomous, and not assembled. Since modular systems are domain-specific computational mechanisms, it follows that they are species of vertical faculties".[74]

To summarise this definition: *domain specificity* means that the operations of the modular system do not cross content domains, and respond to distinct stimulus domains; *innately specified* confirms that the structure of the system is not the result of some learning process, but is genetically determined; *hardwired* defines the system as being associated with a specific neural design; *autonomous*, in a computational sense, refers to the fact that the system does not compete for horizontal resources (memory, etc.); and, finally, the argument for the system *not* being *assembled* from a stock of subsystem processes is propagated by the belief that vertical faculties have a distinct neural structure as outlined in the paragraph above. Fodor clearly states that these attributes do not pertain to *all* cognitive systems, although others since have contested or loosened the author's restrictive clause.[75] Fodor, however, had intended his model to explain only the nature of what he coins *input systems* in a trichotomous functional taxonomy of psychological processes.

This taxonomy is comprised of transducers, input systems, and central processors. Information received by the individual from environmental stimuli is processed in essentially that order. Firstly, *transducer* systems are not computational mechanisms, but rather are those systems that "take proximal stimulations onto more or less precisely co-varying neural signals ... (and) are supposed – at least ideally – to preserve the informational content of their inputs, altering *only* the format in which the information is displayed".[76] For the sake of this argument, I think that it will be useful to understand transducers as our perceptual mechanisms, but, importantly, devoid of any conceptual processing – simply, the neural architecture responsible for *perceiving* sensory stimuli. For this reason it will be important to maintain a distinction between perception and cognition, like Aristotle in his *de Anima*, or as Heidegger states, "Perceiving must be considered totally distinct from the consciousness of a picture".[77] Perception is reliant upon a causal interaction with the external environment, while cognition is not: perception and cognition can therefore vary independently. John Macnamara points out by example that although we may be cognitively aware that the two lines in the Muller-lyer illusion (two equal length lines, but terminated by arrows pointing in opposite directions) are the same length, we do not perceive them as equal.[78]

Input systems, next in the direction of informational flow, have the properties of cognitive modules, as described above. Their functional role is to process the information received from the transducers by *interpreting* it (as opposed to translation) before forwarding it in a useful format to the central processes which formulate conceptual representations and the fixation of beliefs.[79] Fodor maintains that because the nature of central processes must be sensitive to the whole system of belief they must necessarily be able to cross content domains, therefore central processes are not domain specific, and not modular. Rather, he concludes that these non-modular cognitive systems display the qualities assigned to horizontal psychological faculties. Sperber and Wilson in *Relevance* conversely argue that "the deductive processing of information has much of the automatic, unconscious, reflex quality of linguistic decoding and other input processes. What distinguishes the deductive system from the input system is that it applies to conceptual rather than to perceptual representations: that is, to representation with a logical or propositional form".[80]

Following Fodor, input systems, like vertical faculties, are domain specific and respond automatically without any conceptual reflection to their associated stimuli. The most important feature which enables modules to make perceptual identification fast is that they are *informationally encapsulated* by design of their neural architecture. Informational encapsulation restricts communication between domains, or dialogue with other cognitive systems (i.e. central processing systems) by denying access to the information that other systems know about.[81] For example, input systems that interpret information from visual transducers do not share information with those that interpret auditory transduced information, and, likewise, the operations of the input systems are not informed by the central processors, or conceptual representations. Such actions as the focusing and re-focusing of one's awareness on particular perceptual stimuli are instigated in the more interesting realm of the central processes, not in the input systems and certainly not in the transducers. Input systems don't *think*, they just interpret. The processing of perceptual information moves in one direction, from bottom to top, in the direction of the central processes where any real *thinking* goes on.

Jerry Fodor has included language comprehension in the repertoire of input systems, as have other functionalists like Jackendoff and Sperber. Fodor postulates that the automatic triggering of our language comprehension faculty by stimuli is analogous to the reflex action of other perceptual faculties. Such a faculty necessitates a pre-determined structure of linguistic universals, which hallmarks the eccentricity of the stimulus domain. Modularity theory stipulates that the greater the degree of eccentricity exhibited by a domain, the more likely that it is computed by a special-purpose mechanism. In sum, "the eccentricity of the domain rationalises the modularity of the processor and the modularity of the processor goes some way towards explaining how the efficient computation of eccentric domains is possible".[82] This computer-like model of a generalised central processor and specialised input/output devices has increasingly given way to more complex interpretations of the modular composition of the individual input/output mechanisms, as well as an understanding of the central processes being differentiated into "languages of thought" which include propositional, imagistic, kinaesthetic and so on.[83] Jackendoff, upholding a Chomskian position on Universal Grammar, is one who has elaborated upon the complexity of the input/output devices to include a faculty of language communication. He maintains that such a highly specific menu for grammatical organisation constrains the apparent diversity amongst languages to "variations on a theme", which also includes sign languages used by the deaf, and thus demonstrates an innate knowledge for human language capacity, genetically hardwired and automatic.[84]

In *Relevance*, Sperber and Wilson demonstrate that communication based on ostensive-inferential principles can be effected without the use of a code, and from this point proceed to sever any perceived dependency relations between language and communication. They propose that language is requisite for information processing, but not for communication. Verbal communication involves two types of communication processes: "one based on coding and decoding, the other on ostension and inference".[85] The first, usually understood as language, is not autonomous but rather is subservient to the inferential process, suggesting that "linguistic decoding is not so much a part of the comprehension process as something that precedes the real work of understanding", or identifying the speaker's informative intention.[86] It follows that linguistic decoding understood in this sense parallels the functional criteria of input modules, for it is not semantic representations that surface to our awareness as thinking beings but the propositional forms that they explicate which communicate *thought*.

Such determinations about the nature of language, in terms of both comprehension and communication, are particularly important for cognitive studies. A recognition of the modular quality of these cognitive faculties will enable researchers to better frame language as a window onto thought patterns and the nature of cognition. As Jackendoff suggests "since we don't have direct intuitions about the form of thought, we have to use indirect means to get at it. An approach that has proven productive in the past couple of decades is to examine the grammatical form of the language that expresses thought".[87] Before proceeding to its field study application, I will first give a brief account of what Jackendoff has called conceptual semantics.

The idea of a conceptual semantics is meant to ground a theory of "I-concepts", or internalised concepts, which parallel Chomsky's I-language, or language as a body of internally encoded information. Following the hypothesis of an innate Universal Grammar and the theories of syntax, Jackendoff proposes that in order for a finite brain to accommodate an indefinitely large lexicon of concepts, there must be a set of innate generative principles comprising a set of primitives and combination rules that produce concepts.[88] These primitives, like the quarks of particle physics, cannot appear in isolation, but can only be observed in combination as concepts. He resolves that there is an abstract formal algebraic system which determines the major parameters of thought, and which, by design, channels the ways in which experience becomes mentally encoded.[89] In other words, the conceptual semantics of this mental machinery give meaning to sensory patterns, and likely do so irrespective of the sensory modality. These systems are arguably domain specific in their processing, and therefore determined to be examples of what may be referred to in Fodor's modularity sense as 'first order conceptual modules'.

For example, spatial information can be derived from a number of sensory modalities including visual perception, auditory localisation, or the haptic faculty. This would lead us to deduce that the processing of spatial concepts (which inform beliefs of both our first-order conceptual and metarepresentational faculties of thought) cannot be modality specific since it must receive input from any of these indiscriminately. Sensory modalities themselves can be further broken down into different systems, each responsible for processing a specific domain of information: for instance, visual perception involves processing various characteristic of the presented data (including shape of the object(s), colour, texture, motion, and the context of spatial relations) each at different locations in the neocortex.[90] This preliminary processing does not result in concepts, but, rather, manufactures input for the first order conceptual modules. In fact, data processed by the sensory modalities, or input modules, would arguably not be accessible to a conscious awareness (or as objects of intentionality) in this state. For example, the spatial arrangement of neural impulses coming from the retinas, referred to as retinoptic maps, are not "pictures in the head", but must be further processed at a conceptual level to arrive at what Jackendoff has called "visual understanding".[91] Like the things that we say to ourselves "in our minds", visual representations are another means by which conceptual representations become manifest to consciousness[92] and inform the intentionality of our subsequent concepts (beliefs), actions, and behaviour.

The interface between human language and representations is manifest in the way that we use the former to communicate the latter to other humans, thereby acting upon the mutual cognitive environment that we share with one another.[93] A speaker's utterance represents an interpretative expression of their thought, almost always bearing a communicative intention.[94] Words, as utterances, have the power of their associated semantic fields which pick out "cognitive spaces" consisting of a range of values, not simply of a binary or n-ary systems.[95] This broadens the task of utterance interpretation, but also denotes the multiplicity of conceptual representations that a spoken word can signify: 'blue' has a colour semantic field, and an utterance of the word, even in strict reference to colour as opposed to its possible metaphorical implications (if these can in fact be separated), can actuate a broad range of conceptual values in the receiver. With regards to conceptualised spatial representation and the language of space which expresses object description and localisation, there must necessarily be a cognitive process of translation occurring at the interface between these faculties.[96]

With reference to Marr's theory on the processing of visual informa-tion,[97] Jackendoff extends his notion of conceptual semantics to accom-modate spatial representation, and proceeds to show its correlation with spatial language. Marr's hypothesis about the conceptualisation of object description determines that these types of representations are composed of two factors, or primitives, namely: 'i) a set of principles for describing 'general cones' in terms of an axis and cross sections, and ii) a principal for elaborating a main axis [of the object represented], with a subsidiary axis of a particular size and orientation relative to the main axis". Biederman has further elaborated the idea of general cones, postulating that there are thirty-six in total.[98] These cones, or "geons" are the basic units which, combined in structured relations to the axes, compose the array of complex three-dimensional-like forms that can be conceptualised. Any of these geon units may be solid, hollow, or negative, and are detailed in terms of surface features such as colour and texture. The indefinite number of possible object representations supports the vast lexicon of *object names* used in language.[99]

By contrast, the location of figures manipulates significantly less precision in *spatial representation*. Objects are not located in terms of an absolute space, but in reference to other objects and a background which serve to bound the representation. Precision in distance and orientation descriptions relies on culturally produced systems of measurement (quantifiable measurements in distance, angular degrees, etceteras) which "have no special psychological priority", and are thus not component to the representation.[100] Instead, spatial relations are more naturally described in language with the use of prepositions to denote the location of a figure in relation to a reference object, or within a reference region (for example: the lintel is *above* the doorway; the doorway is *inside* the house). Descriptions of the geometric relations between objects specify at most one axis per object, and descriptions in reference to regions additionally specify data about relative distances and directions. In comparison to the rich capacity for descriptions of *what* an object is, prepositions in all languages used to describe *where* an object is are significantly fewer in number than object names. This also equates to an impoverished description of the object geometries, appealing to "the very gross geometry of the coarsest level of representation of the object".[101] It would seem that prepositions filter object descriptions, and retain only the key information concerning the main axis of the object and its bounded-ness.

Jackendoff and Landau conclude that this marked difference in language capacity to describe the *what* of an object and the *where* of an object reflects an innate division in our capacities for spatial representation. This hypothesis is further supported by neurological evidence that description and location are processed at two different areas of the brain.[102] This functional bifurcation in human cognitive design can be explained by a functionalist theory of modularity, in which the *what* and the *where* constitute two separate, domain-specific, first-order conceptual modules. Processed data from both these modules may serve as input to a higher order conceptual faculty which processes cohesive representations which guides our conscious spatial orientation and action in the world. At a philosophical level, further investigation into this may lead us to better understand how space, like time, is in no sense an objective, 'real' entity, but rather a product of our (culturally-informed) consciousness.[103]

The Builders

In summary, there is an evident correlation between the split in the expressive power of language and the anatomical bifurcation for spatial representation. As Jackendoff & Landau correctly suggest in their concluding remarks, linguistic studies can provide a window onto the operations and content encoded by these two cognitive systems.[104]

However, as linguistic analysis provides a key to understanding how cognitive systems encode representations, so too does the study of behaviour and practice reveal the nature of learning skills and the physical manifestation of knowledge. Unlike Jackendoff & Landau's linguistic-based theory which assesses the quantity and characteristics of available vocabulary for *object descriptions* versus that of prepositions for communicating *spatial relations*, I have monitored the 'processes of making' as a window onto human spatial cognition. This methodology proved far more practical in the context of my studies since, most importantly, there was little use of verbal explanation in the teaching-learning processes of the traditional builders, and, as described, inculcation and practice were predominantly bodily. Likewise, in their cognitive study of blacksmiths, Keller & Keller note that language "does not play an essential role in practical forging accomplishments".[105] Rather, "speech is an instrument designed for talking about production, for reflecting on the doing of artist-blacksmithing, but it is not instrumental in the doing itself".[106] In the case of my own studies, I additionally determined that requests for explanations of knowledge from the builders (either in terms of their *'ilm* or *ma'arifa*) resulted in incoherent or incomplete responses as they were not accostomed to verbalising their 'knowing that'. As noted, most often the Master Builders would squeeze their temples with their forefingers and insist that "it is all here, inside our heads". As Bloch observes, it is possible for non-linguistic knowledge to be rendered into language and take the form of explicit discourse, but the knowledge itself changes character in the process.[107] The chief aim of my study has been to understand how conceptual representations and bodily knowledge interface dialectically, one informing the content of the other, and synthesising in the production of practice.

The builder's apprenticeship period served to refine concepts and judgements about space and assembly through training, practice, and inhabiting the processes of *making* space in physical terms. In chapter two on the training of labourers, it was concluded that becoming a Master of the trade means *mastering making*, and before being allowed to *make*, one must discipline the hands, and thus the mind, that will do the making. Once the *usta* deemed the labourer to be in possession of the necessary *aptitudes*, including reason (*'aql*), discipline, and obedience to rank, the labourer may have been granted further tasks and responsibilities that signalled his promotion toward a career in the trade. As illustrated in the case of Saat, Muhammad the Tobacco Chewer, and young Ibrahim al-Maswari, these 'top labourers' also had to display a high degree of motivation in order to win recognition and favour from the *usta*, and had to be tactically competitive in order to retain their positions. The combined insecurity with regard to their status and the verbal abuse received from the senior builders served to reinforce the existing hierarchies of power, as well as to incite a general state of anxiety amongst the precariously positioned apprentices. This anxiety produced an often hostile and curt attitude toward their fellow labourers, reinforcing their competitive temperament, and ultimately reproducing the status-conscious disposition exemplified by the teachers.

In terms of reproducing the trade knowledge, their new duties, working directly under the supervision of the Masters, introduced the young men to the actual processes of building, or *making*, initially through observation and only later through participation. It often involved several months or more of simply providing the *usta* with his tools and materials before the apprenticing labourer was finally handed an adze and asked to carve his first bricks. Meanwhile, the handling and preparation of the building materials combined with close observation of how they are manipulated and employed in the construction, familiarised the labourer with the physical properties and limitations of the building components. Most importantly, this exercise also initiated the structured conceptualisation of the individual building components as being 'assembled' to produce larger components, and these components in turn being manipulated to produce habitable space. This conceptualisation of assembly, beginning with the smallest unit and progressing to the largest, was instilled and reinforced by the schedule of *making* set by the training programme. Tools and instruments involved in the processes of making support the related action-as-knowledge involved, and also constrain or inspire the "formal structures of associated thought and reason".[108] As Keller & Keller argue, an essential connection between conceptual and ecological (environmental) approaches to cognition must be taken into account.[109]

To summarise this aspect of the training programme: the labourer was initiated into the process of *making* with the carving of his first brick. He was then promoted as a *rassah*, still a labourer, to infill the space between the finished surfaces of the walls, notably completing an already bounded space. Next he would be permitted to lay the stone/bricks in erecting the interior surface of the planar walls, which, in the case of the minaret, also involved defining the circumference, and thus the horizontal spatial relation of the wall to the axis of the structure. As illustrated in the earlier section in this chapter on building, the gauging of distances was effected by means of a rope marked by knots and nails along its length, and attached to a central metal pole. After some time the builder, already of senior rank by this point, would be authorised to participate in the assembly of the exterior wall surfaces. Like the interior surface, the contours had to be spatially defined on the horizontal plane in relation to other structural components. However, the inclusion of decorative patterns here interestingly required the builder to extend his notion of spatial relations on the horizontal plane to the vertical plane so that he was able to conceptualise the design as extending beyond the point where he was currently building. Although the decorative patterns were designed and set by the most senior builder(s), this stage nevertheless signalled the young builder's initiation into abstract planning between existing and non-existing components. Once promoted as an accomplished *usta*, the builder would then be responsible for the construction of the spiral stairs, involving complex simultaneous considerations of both horizontal and vertical extensions of the physical components.

Continuous building practice, meaning the regular involvement in the processes of *making*, served to develop a knowledge of the primary building components such as the central column, stairs, and walls, and the manner in which they are assembled to create *space*. At this scale of *making*, one's own body became the principal source of measurement, and by inhabiting the process, reflection upon the body in relation to the physical environment became the principal means of planning spatial relations. Conceptualisations of spatial relations of greater magnitude gradually evolved in conjunction with the repeated practical experience of erecting the minaret. These expanded abstract concepts of spatial organisation enabled the builder to master an understanding of the relations between the larger components of this building type (for example the base, the shaft, the balcony, and the dome), and thereby master the skill which is the hallmark of his trade: namely the expert ability to manipulate the assembly of components for the creation of an entity characterised by a distinctive and 'sophisticated' proportioning of the parts to the whole. Visually satisfying proportion was not something calculated mathematically by the *usta*, but rather, in the words of Ahmad al-Maswari, was recognised as "something that fills my eye".

Plainly, the training process steered the novice from a mastering of *making* toward the mastery of *space*, or, equally, a conceptual mastering of the *object as entity* toward mastery of the *abstract spatial relations between entities*. I will now proceed to explicate an hypothesis of how the builder's training reflected a functional bifurcation in human cognitive design – more specifically between the *what* and the *where* of an object – and how these interface to guide a conscious spatial orientation in both the performance of his tasks and the design of the structure.

To begin with, if an object comprises more than one component (for example a tea pot with a spout, a top, and a handle), then a detailed object description of it implicitly includes a discussion of the spatial relations between the two or more components with the use of prepositions.[110] The object as an entity is therefore defined by the individual component parts arranged in a particular configuration which is defined by a set of spatial relations. It seems reasonable to assert that this argument could apply to any object of more than one dimension by virtue that it has sides, a centre, a top and bottom, a beginning and an end: therefore this would include all physical objects conceptualised by the human mind. What makes the object comprehensible as a single entity, with a single categorical identity (the 'tea pot'), is that its components, and the spatial relations between them, formulate a single cohesive component within a spatial ground, distinguished by material, colour, texture, connection, and maintaining a perimeter skin or boundary which makes it discontinuous, at least conceptually, with other objects in the given context, as well as being conceptually associated with a particular recognised function. Importantly, the conceptualisation of this entity as singular takes precedence over conceptualisations of the individual component parts and the spatial relations which define them.

The young builder, in the process of manufacturing a decorative brick, was engaged in the manipulation of spatial relations through a skilled rearrangement of the object's contours and dimensions. In effect, the spatial relations of the newly made object were being understood and mastered through the *process of making*, and ultimately, the new set of spatial relations were re-conceptualised as a new *object-as-entity*. The increasing magnitude of spatial-relations at each stage in the training, as described above, was mastered in this way. Configurations of bricks were re-conceptualised through the performance of laying them as 'walls', 'decoration' (*naqsh*), or the 'dome' – and so on. Therefore, the *making* was the means both of expressing and of acquiring knowledge, each end of the process informing the content of the other.[111] The traditional builder mastered increasingly complex spatial relations, from the brick to the building, conceptually translating and simplifying these abstract spatial relations into *known objects*. These mastered objects could be manipulated, both conceptually and physically, in an incremental fashion in order to move between scales of different magnitude, from the micro to the macro, over the course of the apprenticeship.

This propensity to conceptually reduce spatial relations into known objects supports the hypothesis of an entropic process of human cognition that guides our understanding of an object from one of abstract spatial relations toward a categorised description of the object-as-entity. These categorical concepts serve as discrete building blocks of knowledge over which the mind of the builder exerts power as a 'maker', manipulating and reconfiguring relations between these objects of knowledge to create new objects of knowledge. This process of manipulating conceptual building blocks is in dialectical relation with the physical processes of making, since, ultimately, the *knowledge* and the *making* are entwined as one and the same.

The Architect

By contrast, the typical training of the professional architect instils first and foremost a mastery of *space*, not *making* (though it should be noted that some schools, and particularly some individuals within those schools, incorporate modelling as an integral part of the design process). Drawing, and now increasingly computer-aided drawing, is the principal means by which the majority of architects communicate their ideas about building to others, and the medium through which they entertain a discourse between their own objectified ideas and conceptual knowledge. Like the builder for whom knowledge is expressed through the *making* (or actually *building*), the architect's knowledge becomes objectified through the process of drawing. Drawings and imagery-based concepts inform one another through this process, thereby empowering the architect to manipulate and evolve his concepts of essentially spatial relations. Generally, through the process of *drawing* the architect produces knowledge about *a* building, whereas the traditional builder, through the process of *making*, produces knowledge about *building*.

Having had the opportunity to assist in the department of architecture at Sana'a University as an occasional design critic for the student projects, I was able to compare and contrast the course curriculum and design studio approach with my own undergraduate studies in the field, and with my brief teaching experience in a department of architecture in Northern Nigeria. In theory, the training in Sana'a, like that in Nigeria, was very heavily modelled after Western precedents, and the students were streamed from engineering backgrounds. However, in practice, it has been burdensome for the staff and administration to provide a quality standard of education with the loss of Kuwaiti financial support for the university since the Gulf War. Nevertheless, design was taught and practised in a way familiar to me as an exercise in qualifying and quantifying spatial relations between imagined solids and voids.

When I began my studies in architecture at McGill University in Montreal, one of the first assignments was for each student to draw the floor plans of his/her house, first from memory, then from a measured survey, in order that we compare and contrast our imagined spatial relations of the places we inhabited with the actual ones. The devising of convincingly functional, and sometimes seductively intriguing, plans was to remain the hallmark of our education. In the latter years of the programme we were actively encouraged by our professors to pay special consideration to the materials and construction assemblies in our building designs. The drawing of building sections and wall assemblies took us conceptually closer to the materials and assemblies we proposed, but the drawings were still mainly a conveyance of spatial relations, albeit at the micro level of the building planning. Even the detailed working drawing for, say, a special stone cut communicates spatial relations and dimensions through its drawn lines and mathematics, not the stone itself. Drawing ultimately remains a separate enterprise from the *making* on the building site. Translation between the two requires a (complex) transformation of knowledge. Arguably those involved in the processes of model making are expressing and formulating knowledge through *making*, but, again, unless the materials and scale equate to those of the proposed building, then the process of making remains different from that of the builders. Producing a massing model from balsa wood or bent copper sheets does not involve the same 'line of thinking' as assembling a structure with basic building units (which themselves may be modified, or 'made') to create larger 'objects', which in turn become the units of yet larger 'objects' in the building assembly.

To conclude, in the context of this discussion, the most noteworthy difference between the training of an architect and that of the Sana'ani traditional builder is in the 'lines of thinking' imposed by their respective teaching-learning processes, and the resulting reliance upon different domains of knowledge. The traditional builder, through the process of *making*, reliant upon his sensorimotor domain which both produces his skilled performance and his *knowledge about building*, begins his training by mastering the basic unit of the brick. Each stage in his development is about the mastery of increasing scales of spatial relations, and the consequent formation of the object-as-entity. In other words, the builder's knowledge moves from the micro to the macro scale, or from the brick to the building, mastering his understanding of spatial relations by transforming them into *objects of knowledge* through the physical process of making *concrete* objects that can be *known*. Conversely, the training of the architect, as outlined briefly above, first inculcates a knowledge of spatial relations at the scale of the building plan, and moves the architect's knowledge, predominantly through the process of *drawing* reliant upon imagery-based conceptualisation (and the skill of drawing itself upon the sensorimotor domain), towards an understanding of the 'particular' in terms of drawn details of the building's component parts. In other words, as opposed to the 'micro to macro' direction of the builder's training in making, the architect's knowledge about *a* building, and more particularly about *abstract spatial relations*, evolved inversely, from the macro to the micro scale. Subsequently, through their respective training, the builder becomes a "Master of Making" and the architect, a "Master of Space".

Concluding Remarks

A Sana'ani architect shared his grave concerns with me about the future of traditional building in his country, especially in the capital which has undergone sweeping social and physical changes in the last few decades. He complained that "young people are no longer interested in this work and the ageing knowledgeable Masters are retiring from the trade". To remedy this apparent predicament, the same architect proposed that a craft school be set up in order to train a young generation of builders about using traditional materials and methods, and thus preserve this expert knowledge. His proposition echoed a similar bid put forth by the Italian organisation, Studio Quaroni, for a Sana'ani masonry school in their detailed report in 1984;[112] as well as by Lewcock in his strategy for conservation.[113] The architect's 'consternation' about a disappearing tradition of building crafts was common on two accounts: firstly, many Sana'anis didn't realise to what extent the labour force, and very often the Master Builders themselves, came from outlying rural regions and not from the brigades of children that grew up playing on the streets of their urban neighbourhoods who would, in the majority of cases, become traders, business(wo)men, civil servants, etcetera; secondly, as an architect serving a Yemeni public, his expertise continued to play second fiddle to the authority of the *usta* in the popular mind of the people. I often suspected that members of the architectural community somehow hoped that the traditional builders were in fact disappearing in order that they might eventually increase both the number of commissions they received and the status of their own professional power. Professionally trained architects were a recent, and arguably superfluous, addition to this nation of remarkable builders.

More importantly, what the architect (and Studio Quaroni et al.) was failing to consider in his proposal to replace the supposedly waning apprenticeship system with a trade school was that, because of its very nature, it would be nearly impossible to institutionalise the processes of this distinctive education. The training of the builder is deeply rooted in the formation of his person and not simply about teaching skills necessary for craft reproduction. As illustrated in the last chapter in which the builder's training was compared with early religious socialisation, the formation of discipline in both mind and body is crucial for producing an individual who is capable of acting, thinking and understanding within the framework of their vocation. Once the builder transcends his status as a common labourer to become an apprentice, the increased insecurity and the mistreatment by those senior in rank serves to procure a disposition of keen competitiveness toward his fellow workers, and ultimately, with achievement, a sense of superiority, thus reproducing the existing hierarchy of rank. Discipline, resolute attitude of superiority, and skilled performance, are integral qualities of the *usta*, and it is the inculcation of all three during the course of the training that defined his expert status.

Trade schools, like the one described for training Quebec carpenters, objectify knowledge (as *'ilm*, arguably with little or no sense of *ma'arifa*) through language and drawing in order that this be transmitted efficiently and systematically to the students. The process of objectifying knowledge renders concepts and ideas amenable to public scrutiny, to being discussed, challenged and refuted, and possibly changed. Such a process would undermine what I have identified as the 'resolute attitude of superiority' of the Master Builder, and thus his real power as perceived by the general public. This issue will be addressed more fully in the next chapter. What is noteworthy here is that, in comparison with institutionalised education programmes, the use of propositional forms of information and explanation in the training of a building apprentice was almost negligible. This absence made it difficult for an outside agent, and, as I have stated, even for the young builder himself, to question *what* he knows and *how* he knows. His knowledge was sedimented in his performance: in the *making*. Therefore, it was the novice's physical involvement in the *processes of making* which defined his role as an 'apprentice', and not the local use of any popular lexical term to rigidly classify his position. It was the builder's body that was treated as memory, and "the principles em-bodied in this way are placed beyond the grasp of consciousness, and hence cannot be touched by voluntary, deliberate transformation, [and] cannot even be made explicit".[114]

A comparative analysis of teaching-learning methods was also made between the training of architects and of traditional builders, and the resultant types of knowledge regarding *space* and *making* were considered. As opposed to the apprenticeship of a traditional Sanaʿani builder which focusess on the mastering of making, the education of an architect stresses a mastering of space. The architect is one chiefly practised in devising planning concepts which remain ideational until executed by a construction team. Therefore processes of conceptual design and of construction remain separate, fully reliant on the mediation properties of the architectural drawings, which involve not only a translation, but a transformation of knowledge. In contrast, for the Sanaʿani traditional builder it is the process of making, rather than drawing, which serves as the mediating tool between idea and physical structure as product. I have argued that it is the inherent properties of the mediating tools which categorise each distinct type of training process. Furthermore, I have demonstrated that the dominant employment of each mediating tool, the drawing or the making, in the teaching-learning process results in different structures of the individual trainee's spatial cognition. In brief, the architectural training, and in particular the process of thinking-through-drawing, fosters skills for the conceptual mapping of vectorial space incumbent on abstract systems of measurement and manipulated by a mathematical language; whereas the apprenticeship system fosters a conceptual mastery of spatial relations through direct association with the particular physical entity, namely the *making*.

Chapter Four

COMPLETING THE DOME
The Master Builders

The movements of nature which we define spatio-temporally, these movements do not flow off 'in time' as 'in' a channel. They are as such completely time-free. They are encountered 'in' time only insofar as their being is discovered as pure nature. They are encountered 'in' the time which we ourselves are.

Martin Heidegger[1]

Chapter three focused on the apprentices' training under the guidance of their masters, and on the trials confronted by these young men in their efforts to attain and preserve their positions amongst their fellow labourers. The 'slipperiness' of lexical categories used to define the apprentice's position was seen to be a reflection of the fluid quality of the power structure below the level of *usta*. It was determined that the novice's participation in the *processes of making*, and not a fixed classificatory title, is what defined his role as an apprentice to the Master. A mastering of *making* was the distinguishing feature of the traditional builder's education, and this was shown to contrast with the typical education of an architect which emphasises the mastering of *space*. A modular theory of mind was employed to postulate a bifurcation in spatial cognition between the conceptualisation of *objects-as-entities* and that of the *spatial relations-between-entities*, and this was subsequently supported by the comparison between the teaching-learning processes of Sanaʿa's traditional builders and the architects who were studying at the university.

Again, this chapter has been divided into two main sections. The first will describe the final stages of minaret construction, namely the building of the dome and the application of lime plaster used to highlight the decorative brickwork, the perforations, and the contour lines of the tower. The common styles of domes will be discussed, and an investigation will be made into the lexical source of the elegant fluted-dome type, *muthallajah*, which caps many of the historically significant minarets of Sanaʿa. Once the structure has been completed, the bronze *hilal*, or decorative finial surmounted by a crescent moon, is secured onto the *qasabah*, or metal axial post, which pierces the dome's crest. Metaphorically, I have linked these final building stages with the final stages in the builder's career as an *usta*, or Master Builder, which represents the theme of this chapter. In this first section, the four *asatiyyah* of the Bayt al-Maswari and the processes involved in their design and construction decisions will be introduced.

The following section will begin with the introduction of the al-Maswari brothers, and will demonstrate the link between the state of their current trade knowledge with the instruction they received from their father, and with their own involvement in the transfer of this expertise to the next generation. In the next sub-section entitled *Earning an Expert's Expertise*, the on-going acquisition of expert knowledge by the younger *usta* under the tutelage of his senior Master will be considered, and the politics implicated in gaining further autonomy from the Head Master will be discussed. The following sub-section will investigate the builder's conceptual processes of design, with particular reference to the creation of decorative brick patterns and the occasional use of drawings to objectify ideas. The nature of the Master craftsman's 'traditional' knowledge will be explored and the notion of *intentionality* shall be introduced before proceeding to a more complete discussion on this topic.

The topic of *Expertise and Intentionality* has been divided into two parts. Firstly, I aim to develop a useful conceptual model of intentionality based on Heidegger's theory of time consciousness, and supported by neuro-philosophy, in order to frame subsequent discussions around the intentionality of both project patrons and the Master Builders. The multiple motivations for *sadaqah*, or charitable endowments, shall be investigated, and the effects of craftsman-patron relations on the intentionality of the chief builders will be discussed. The level of intentionality achieved by the Head Master Builder is compared with *ihsan*, the ultimate dimension of Islamic spirituality. It shall be determined that the 'balance' between the conceptual forces of imagination and reason, wrought by the *usta*'s highly focused and disciplined concentration on his subject, is the *sine qua non* of his expert status. In conclusion, the contention for power over the expert discourse between architects and traditional builders will be examined, and some of the factors necessitating the re-negotiation of the power share are to be considered.

I About Building: a view from up top

"It looked like an old woman bent over, like an old hunchbacked woman (*hadba'*)" complained Ahmad. "With such a stunted brick shaft atop that towering stone base (*murabba' hajar*)", he continued, "she was all legs and no body". Shortly after resuming construction of the octagonal shaft above the level of the balcony, both Ahmad and his brother Muhammad decided that the height of the minaret for the Masjid 'Addil would need to be higher than originally planned. The project budget was calculated for the erection of a forty-four meter high structure, but, as the building progressed from the ground up, it was realised that capping the minaret at that height would result in an aesthetically displeasing tower of awkward proportions. The two senior al-Maswaris informed their patron, the members of the Bayt Fahim, that they would add ten meters to the stipulated height. "If further funding could be allocated, it would be much welcomed" they added, but otherwise the Bayt al-Maswari themselves would finance the corrective measures "in order to preserve the memory of 'Ali Said", their deceased father and mentor, and to "maintain the reputation" of their House as excellent builders (see figures 9 & 10). This devotion to their father and to the perpetuation of the family reputation was publicly expressed in stone inscriptions placed above the doorways of several minaret projects, reading: "The execution of the work of the minaret was by the sons of 'Ali Said al-Maswari".

From the balcony level, thirty-one more stairs were joined to the staircase whose spiral boasted a total of one hundred and seventy-five risers upon reaching the final, and interior, prayer-calling platform. The minaret's outer brick shell, reduced in thickness from four bricks to three from the balcony upwards, maintained its steady inner circumference hugging the stair shaft, and an octagonal exterior plan built without any further geometric transformations (see figure 11 and 15). From the level of the final stair, it would take one more month to complete the nearly eight meters of vertical construction required to reach the crest of the dome. The termination of the staircase which, importantly, had provided a means of access to the top of the walls in this structure being built from the inside out, meant that the construction process itself became more complicated. Alternative forms of interior scaffolding were required in order for the builders to reach their stations on the ever-rising walls. The central stone column, like the stairs, was terminated at the level of the last calling platform. Above it was built a narrow brick wall, one *ajurr* brick wide and the length of the column radius, located to the east side of the axial metal post, or *qasabah*. This brick wall, effectively a small pier built atop the larger column, was built upward to a point below the inner surface of the dome, and served primarily to support a series of working platforms which were constructed as needed.

Large apertures, about two meters in height, were introduced into each of the eight vertical planes of the shaft at the level of the highest calling platform. During the construction of the eight pier-like corners which frame the apertures, pairs of sturdy tamarisk branches were set into the brickwork, spanning the width of these openings. The branches were used to support wooden planks which in turn spanned between opposite openings and across the diameter of the shaft. When the branches were no

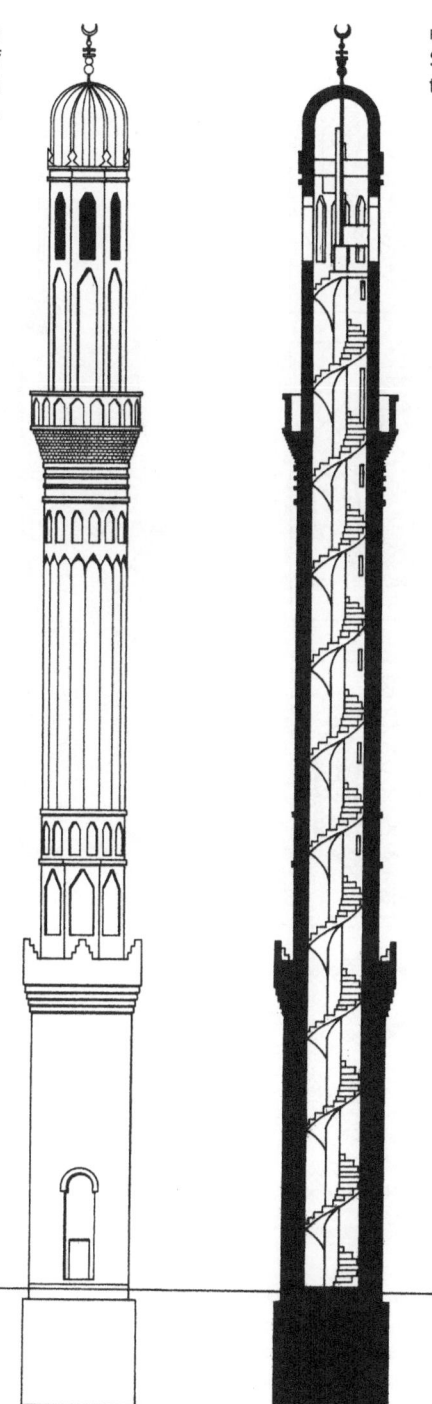

FIGURE 9
Elevation drawing of
the completed 'Addil
minaret.

FIGURE 10
Section drawing through
the 'Addil minaret.

187

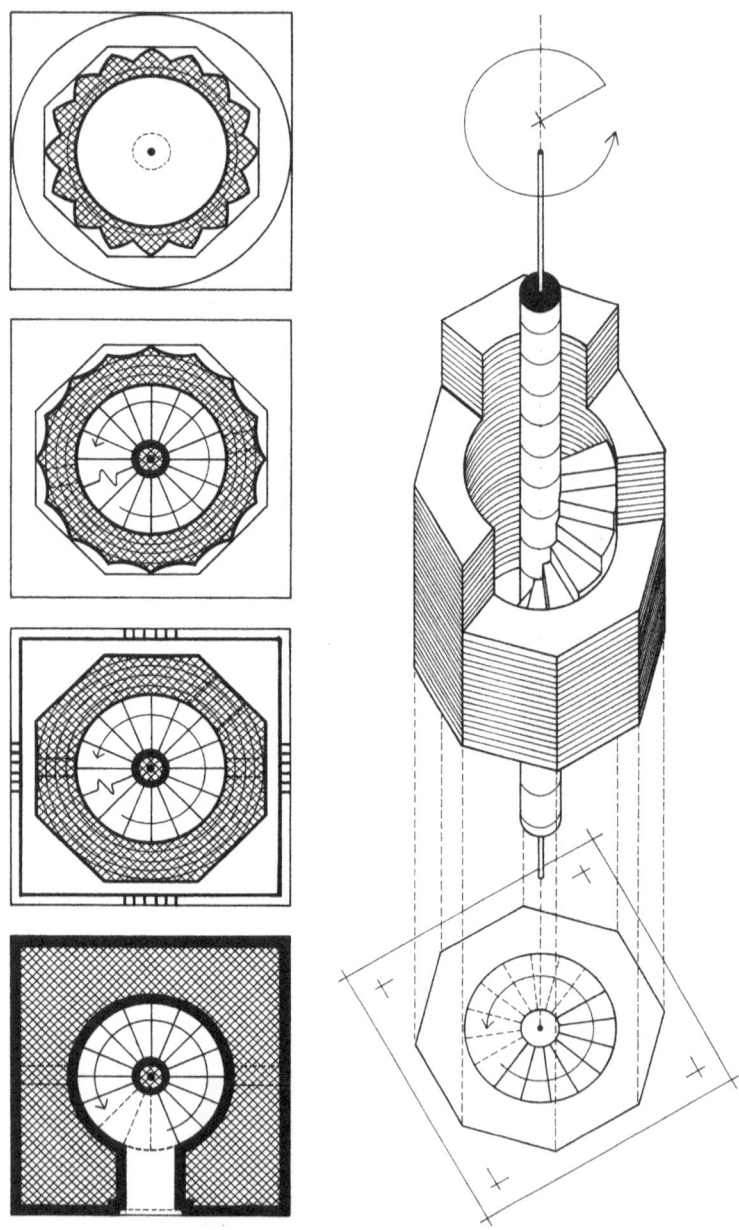

FIGURE 11 Axonometric drawing of the octagonal brick tower
with the axial metal rod passing from the bottom to the top
through the centre of the column. And four superimposed
plans of (from bottom to top) the minaret base; the octagonal
shaft; the fluted shaft; the fluted dome.

longer needed, they could be sawed away. Majid, Muhammad's son and the third *usta*, took the credit for this technical innovation and conceitedly claimed that his father and uncle would never have arrived at such a novel idea. The Master Builders balanced themselves precariously on the make-shift scaffolding in order to lay bricks, and their apprentice, centrally positioned where the planks crossed each other in a star-like pattern, distributed their materials. The materials were handed on to the apprentice by a second labourer who was somewhat dangerously located below the bowing planks of old timber amongst the stock piles of materials on the calling platform.

The apertures were completed at a height of nearly two meters in flat-topped corbel arch configurations and capped with lintels of sturdy tamarisk branches. Once the structure was completed, four giant bell-shaped loudspeakers were hoisted to the top by ropes and installed in four of the eight windows. Above the openings, the eight planes of the exterior wall were terminated one meter higher with a decorative cornice in relief brickwork. The eight corners of the planes were each vertically accentuated with decorative finials built in carved bricks, like jewelled points crowning the base of the high dome. Inside the structure, a series of platforms, five in total, like giant stairs varying in rise from one to two meters and ascending in the counter-clockwise direction, were built to ultimately give working access to the level of the dome. The first three, rising a total height of slightly more than two and a half meters above the calling platform, made a total turn of one hundred and eighty degrees. The fourth platform, about a half meter higher, was located inside the base of the dome and covered a floor plate of two hundred and seventy degrees, or three quarters of the circular area, and the fifth, another meter higher, covered the remaining ninety degrees of the circular plate. All were constructed using timber beams and brick risers which spanned the radius of the open shaft between the narrow brick pier resting on top of the column and the minaret's outer wall.

With the completion of the tower walls, standing now at an impressive height of over fifty meters above the forecourt of the 'Addil mosque, it came time to construct the dome. There are two prevalent types of domes used to cap traditional Sana'ani minarets: a smooth-surfaced dome which may be either spherical, like that of the recently reconstructed Masjid al-Shahidayn, or slightly bulbous like those of the city's Great Mosque, the first of which was built in the 9th century AD and heavily restored in the 13th century; or fluted, like that of the Masjid al-Abhar (1374–75 AD) or the al-Madrasah Mosque built in the first half of the 16th century, which look rather like the ribbed bulbous element in the centre of a typical glass orange-juicer. The Bayt al-Maswari had constructed both types for their numerous past commissions and were planning to build a smooth spherical dome for the 'Addil minaret. As an anthropologist, as well as a 'mere labourer',

I normally refrained from contributing any input into the decision-making processes at the building site, but it must be admitted that, as an architect, it was difficult not to suggest what my aesthetic preferences were concerning the cupola. I had been much impressed by the grace of the fluted domes in Sana'a and, more so, I was technically interested in how they were constructed. In a subtle, but perhaps persistent way, I expressed admiration for those fluted domes which the al-Maswaris had built in the past, and exhibited mild disappointment when they repeatedly insisted that 'Addil would have a smooth, round cupola, or *qubbah*.

Before construction on the dome commenced, I spent a week away from Sana'a assisting a fellow colleague in measuring some of the crumbling building remains of wind-swept Mocha for her research, and allowing my aching body and dust-filled lungs to recuperate. I returned on the following Saturday feeling rejuvenated, and met Muhammad, the head usta, enjoying the warmth of the early morning sunshine perched on a stone door stoop across the road from the minaret. He greeted me warmly and, with a sheepish grin, looked away from me, sweeping his attention upwards toward the top of the tower and inviting my eyes to follow his gaze. "What do you think?" he asked, "Quite nice, yes?". The base of the dome, already visible from the street below, was fluted. I couldn't contain my sense of satisfaction. In my mind this was the most elegant way to terminate this slender proportioned edifice. I asked Muhammad what he called this type of dome to which he responded, "*Muthallajah*" (see figures 12–14). I scribbled this down in my notebook and confirmed my spelling with him. That evening, having had no success in finding a logical root of *muthallajah* (*th.l.j*) which could be applied to architecture in my Arabic dictionaries, I began what was to become an intriguing lexical investigation, the finds of which I will briefly describe here.

Returning to the site the following day I approached the *asatiyyah* during the breakfast break, all squatting with their tea and *ruwtee* bread in the mosque forecourt, to once again confirm my spelling and to ask for their definition of this strange word. Muhammad took a sip from the condensed milk tin which he and the others used as a tea cup, and handed it onto his brother Ahmad before replying. "*Muthallajah* means to be like melted ice" he said slowly and confidently. "Yes, like melted ice" Ahmad chimed in with a mouth full of bread. The two younger Master Builders remained quiet, and 'Abdullah al-Samawi looked away disinterestedly toward the gate leading out into the street. Though I knew that *thalj* means either snow or ice, I was puzzled. Muhammad continued, "If you leave a lump of ice in the sun, it will melt, and the water running down the sides of the ice will form crevices like those on the dome. See?". His brother shook his head in accord. I conceded politely to their explanation, but in fact I was not convinced by it and had my suspicions as to whether they had ever seen a melting block of ice.

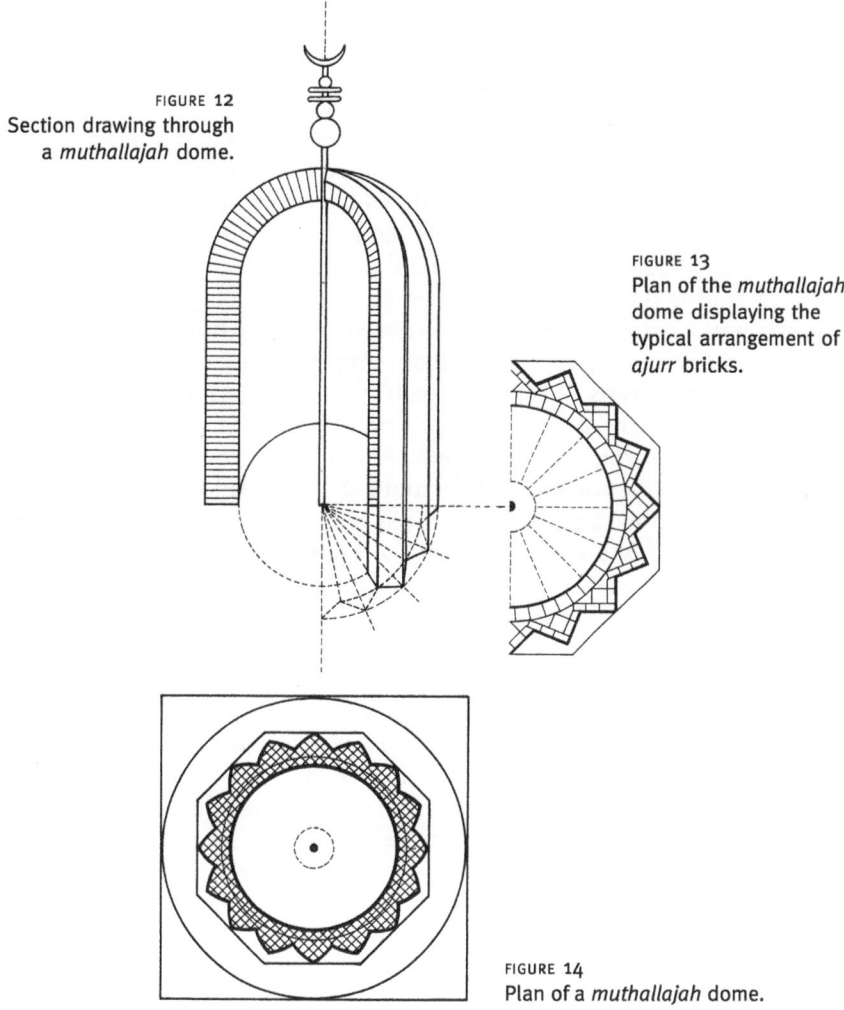

FIGURE 12
Section drawing through
a *muthallajah* dome.

FIGURE 13
Plan of the *muthallajah*
dome displaying the
typical arrangement of
ajurr bricks.

FIGURE 14
Plan of a *muthallajah* dome.

During a *qat* chewing session later that week, I probed for answers from a learned Sanaʿani friend who had a keen interest in both Arabic and English semantics, and spoke the two languages masterfully. He had never heard the word *muthallajah* before, but also knew of no alternative term for describing these classic Sanaʿani fluted cupolas. After giving some thought to the matter, arguably made more lucid from the effects of the *qat* stored in his cheek, he speculated that the word may be a corruption of *mufallajah*, derived from the root *falj*, meaning a 'crack, split, or crevice'.[2] He went on to explain that it is fairly common for the Arabic letters '*th*' and '*f*' to be swapped in the spoken Sanaʿani dialect, and native speakers are seldom aware of this fact. When I next approached the builders with this alternative definition with the aim of eliciting further discussion on the matter, none were familiar with the classical term *falj*. The idea of a phonetic corruption did, however, set some wheels in motion, and Majid later came to me proclaiming that "*muthallajah* is a mispronunciation of *muthallath*", meaning 'triple, threefold or triangular'.[3] "Look!" he said, directing my attention to his hands which he held out lengthways at an angle to one another, "Just like the dome", referring to the profile of the dome's individual triangular-shaped fillets.

My Arabic tutor suggested *thalm* as a possible origin to the mysterious word, and one which also may have been bastardised from an original form, *muthallamah*, to *muthallajah* over the course of time. The root, *thalm*, which means a 'nick or notch',[4] could describe the crevices of the dome's flutings, and seemed as viable an interpretation as the last two I had collected. None yet, however, were certain. Some months later, an English friend who is an excellent Arabist and has long resided in Sana'a, rendered what I have considered the most plausible explanation. In the autobiography of Qadi Muhammad al-Alewa (volume 2: page 36), he discovered the word *mufallajah* in the Arabic text, the same word first suggested to me during the *qat* chew. In the context of the passage, the author employed the word in reference to the ribbed periphery of a round, plate-like alabaster oil lamp once common in Yemen. The similarities between the ribbed form of the lamp and the flutings of the dome strongly suggests that the word *mufallajah* is indeed an adjective borrowed to describe the architecture. Therefore *muthallajah*, with a "th" rather than a "f", would be a phonetic corruption of the original word. Also, as these lamps are no longer popularly used since all houses in the capital are supplied with electricity, the original source of the word has not surprisingly been forgotten by the builders. In fact, some months later, this hypothesis was strengthened when one of the Master Builders referred to the scalloping of the exterior face of the brick walls of the al-Ihsan minaret in Madinat al-Asbahi as *muthallajah*, like the dome. In plan, a cut through this section of the minaret tower convincingly resembles the plan view of a typical Yemeni alabaster oil lamp (see figure 16).

Typically, the *muthallajah* domes constructed by the Bayt al-Maswari have sixteen fillets, or ribs, governed by the 'multiple-of-four' geometries of the transforming structure: two flutings incised above each of the eight planar faces of the wall below, and aligned with the sixteen scallops situated further down the shaft below the level of the balcony. Sixteen divisions is not a rule for Sana'ani *muthallajah* domes, as some may have only eight fillets, such as that of Masjid al-Abhar, Masjid Salah al-Din, or Masjid Dawd (all within the walls of Old Sana'a). Other minarets less skilfully constructed appear to have an arbitrary number of components, likely haphazardly dictated by the availability of space rather than by design. The dome of the

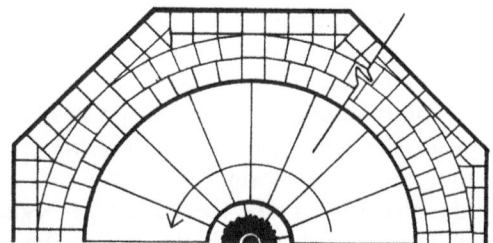

FIGURE 15 Plan of the octagonal tower above the calling platform displaying the typical arrangement of *ajurr* bricks.

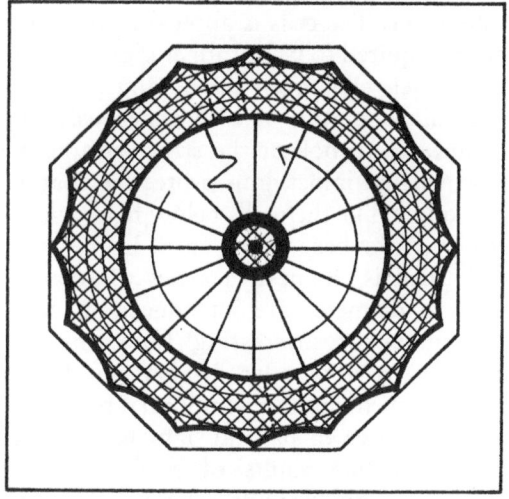

FIGURE 16 Plan of the scalloped, or *muthallajah*, section of the brick tower displaying a similar configuration to a South Arabian alabaster oil lamp.

'Addil minaret was constructed entirely in *ajurr* brick, and the consistency of its inner spherical surface and the contours of the fluted exterior were regulated once again by points fixed on a radial rope attached to the *qasabah*, or axial metal post. It can be recalled that for the construction of the walls, the rope, with fixed radial points measuring out contours on the horizontal plane, was slid up along the post in accordance with the rising height of the tower. In constructing the curvature of the dome the rope was fixed vertically at a point on the *qasabah* and was used to measure the hemispherical contours in all three dimensions. This same method of dome construction was observed in the erection of a public bathhouse in the vicinity of the al-Ihsan Mosque, where, rather than an axial post, a rope was fixed to the centre of a temporary wood beam that spanned the circular base of the cupola supported on squinches.

193

By experience, the al-Maswaris have learned the merits of vertically elevating the base of the dome before introducing the curvature in order that the cupola be adequately viewed from the street below. The dome of their first minaret, commissioned for the Masjid al-Husayni around 1980, was built without an elevated base and is therefore perceptible only as a white sliver floating above the tower walls, noticeably out of proportion with the rest of the structure. Once the curvature begins, a maximum of four brick courses can be laid in a single day as there is no scaffolding system to support the inward sloping sides of the dome. It is erected as a self-supporting structure, and a stronger, faster drying mix of quicklime mortar, or *guss*, is used to bond the bricks rather than the normal mortar, or *khalta*, used in the walls. The *guss* is applied to the brick surface by hand, and the builders are required to wear rubber gloves and a full-length overall in order to protect their skin from the acidity of the lime. As the dome proceeds, the quicklime mixture is also generously applied on both the inner and outer surfaces to bolster the structure's homogeneity and, very importantly, to provide an impervious protective coating.

The shell of the dome, like the walls below from which it springs, is three *ajurr* bricks thick from the base to the apex. In the trough of each fluting, where the dome's shell is thinnest, a series of small perforations about fifteen centimetres in diameter are introduced in vertical arrangements. These serve a number of interesting functions. Firstly, during the construction itself, narrow wooden beams can be inserted into these perforations which in turn support temporary platforms for the builders. They also provide a source of natural light to the interior of the dome, reminiscent of the star-like vaults of a Turkish bath. Perhaps most importantly, the perforations provide grips for the hands and feet of the builder who undertakes the daring task of scaling the exterior surface to install the brass crescent moon, or *hilal*, or for those commissioned to apply the decorative *guss* on the exterior or to perform future maintenance work. A small slit opening along one of the flutes is made halfway up the height of the dome, very often on the south side, and just large enough for a worker to pass through in order to gain access to the exterior. I can report with some relief that the opening on the 'Addil minaret was too narrow for me to exit and therefore I safely avoided having to manoeuvre myself along a steeply curved surface over fifty meters in the air.

Progressively there was less and less room at the top, and the number of Master Builders constructing the dome was reduced to two, Ahmad and 'Abdullah al-Samawi. They were assisted by Saat who prepared the quicklime mortar and bricks, and distributed any necessary tools. Below Saat, between where he was located on a platform in the dome and the top of the stairs, were two other labourers responsible for passing on heavy bags of lime and bricks from the stockpiles created on the last calling platform. The task of moving materials upward was made more difficult by the steep and awkward series of levels that had now replaced the staircase. Once a sufficient reserve of bricks was secured at the top, there was no need to employ a large battery of labourers to man the winding staircase or to prepare materials in the mosque forecourt, and their numbers were reduced to five. Occasionally when Muhammad al-Maswari would pass his labourers who were standing about in the staircase waiting for supplies to be passed along, he would shout at them, calling them *"Humur!"* (donkeys), and reprimand them for their laziness. Muhammad sensed a certain euphoria brewing as the project came to a close, and he knew that without constant disciplining the labourers were likely to neglect their outstanding tasks altogether. This was in fact true, and small groups of labourers increasingly began to congregate together and lounge about on the stairs chatting and smoking cigarettes. This provided a wonderful social opportunity to tell jokes and exchange gossip, but, as Muhammad had suspected, duties became a less pressing issue on the minds of the builders.

Several labourers were transferred from their position in the stairs to work alongside Majid and Muhammad al-Maswari on the completion of the balcony wall. This presented an exciting experience for many of the young and ambitious builders, allowing them to work more closely with the *asatiyyah* and to hone some of the skills that would be required of them as future apprentices. In some instances this involved the carving of bricks used in the decorative relief panels, or directly assisting the Masters by distributing their tools and preparing materials. More commonly their tasks comprised rendering the exterior brick surface of the balcony wall by rubbing it smooth with wet, runny mortar and a half-brick held like a brush in order to achieve a homogenous earth-red surface, as described in the last chapter. The *ajurr* bricks are brittle and break quite easily. This is particularly dangerous when performing the latter mentioned task on an exterior surface some forty meters above the street. In one instance the brick that I was using to rub the wall crumbled in my grip and fell into the street below, exploding on impact and leaving a red star-burst pattern on the asphalt. It was very good fortune indeed that there was no one passing at that given moment.

The balcony wall was divided into thirty-six sections around its circumference, comprised of eighteen plane-faced sections alternating with eighteen decorative brick relief panels. Initially these sections were laid out around the perimeter by Ahmad using a stick cut to a fixed length. This method required several trials and several adjustments to the length of the stick in order that all thirty-six divisions fit precisely along the outer contour of the balcony platform. Interestingly, and with no explanation, the divisions were not made to conform or align with the other transforming geometries of the tower. To my mind, it would have been more consistent with the mathematics of the structure if thirty-two sections would have been measured out and aligned with the sixteen scallops on the shaft below.

195

Nevertheless, once these were determined and marked out, one *usta* would lay the brick courses of the plane-faced panels, and the second would follow after him setting the decorative patterns, both moving around the platform typically in the counter-clockwise direction. A copper section of piping was installed at the base of the wall to allow for drainage, and a concrete floor of about eight centimetres thick was set in order to protect the cantilevered brick structure below against the infiltration of rainwater. While hauling up endless buckets of concrete, one of my co-workers with aching arms commented on how he wished that we had a crane at the minaret like the one he had seen at a modern building site on Zubeiry Street. I reminded him that *he* wouldn't be here if the *crane* was. "Oh yes! True", he said with an amused smile.

Meanwhile, the sides of the dome were being steadily closed in upon themselves, and now only 'Abdullah and Saat, his apprentice, remained working in the confined space at the top. During my visits there, crouching with Saat on the highest platform, pressed up close against the underside of the white plastered dome and feeling the draught of cold November air, I had the distinct impression of being inside an igloo. I strongly suspect that my clearly Canadian analogy was not shared by my Yemeni colleagues. The axial metal post had been extended to the exterior beyond the point where the vertex of the dome would ultimately be, and it would eventually support the brass *hilal*. An oculi of about one meter in diameter was left around the *qasabah*, and from this point onwards 'Abdullah worked from the exterior, climbing onto the dome through the top and setting the shell's innermost layer of bricks. He was no longer setting entire courses around the circumference, but rather was bridging small sections between the completed edge of the ring and the central post. Progressively, as the *guss* dried and the bonds hardened, he set the weight of the outer two layers of brick upon the first. After each small section was completed, he re-entered the interior space of the cupola and applied a coating of the quicklime mixture to the inner surface by hand.

As the sixteen fillets and flutes closed in toward the apex, it was impossible to maintain their perfect geometries, and the arrangement of bricks, now at a level which would be imperceptible from the street below, were set in more haphazard arrangements with the main aim of completing a solid hemispherical structure. In the final stages, the top of the dome could only be accessed via the slit opening on the south side, and wetted bricks and quicklime mortar were carefully passed up from here to the builder on top completing the process. Finally, with the dome structure complete, all that remained to be done was the installation of the *hilal* and the highlighting of the relief brickwork with *guss* by a team of specialists in this notoriously dangerous trade.

Initially I was told that the *hilal nahas asfar*, or brass crescent moon, to be placed on the pinnacle of the 'Addil minaret would be imported from a

PLATE 47
A *hilal* for sale in Cairo.

workshop in Damascus at a price of 250 000 Yemeni riyals (approximately two thousand American dollars). Although *ahalil* (plural of *hilal*) are locally manufactured in the *suq*, the indigenous models are considered to be of lower quality than those produced in Cairo, Damascus, or India (see plate 47). The imported quality also lends status, I was told, to the patron of the religious endowment. In the case of the al-Maswari clients, all were notably wealthy, usually with a high ranking status in the business community like the Bayt Fahim. In all their minaret constructions to date, the Bayt al-Maswari have employed the more standard and traditional crescent moon motif. There is also, however, a tradition in Sana'a of using a finial surmounted by the brass mould of a pigeon, such as on the minarets of the al-Madrasah and the Salah al-Din Mosques, two of the earliest prayer calling towers of the Old City. After a wait of several months, the *hilal* was finally imported from India, not Damascus, and a garish flashing pink light was installed on the top of a metal pole in the interim to visually designate the now fully-operational mode of the

197

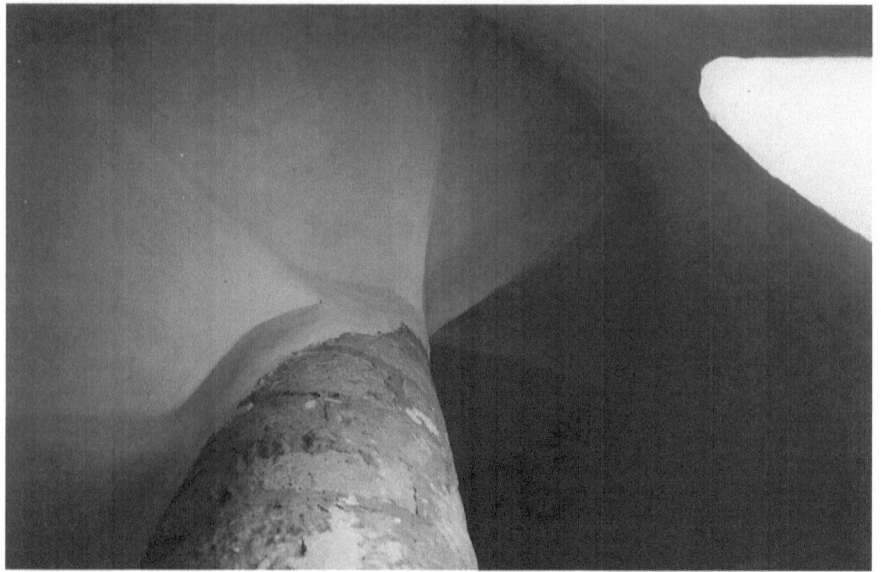

PLATE 48 A view of the underside of the spiral staircase after it
had been plastered.

tower. The loudspeakers had already been crackling their message across
the quarter since a slightly earlier date.

In June of the following year, more than six months after the
completion by the builders, a small team of *guss* workers specialising in
the plastering of the decorative brick relief and the interior surfaces
arrived to execute the final stages of the ʿAddil minaret project. The
Master of the team was a man named Muhammad al-Haimi, and he was
assisted by his two adolescent nephews, the younger of which was keenly
interested in collecting giant locusts that flew into the top of the tower and
stuffing them into a plastic bag to take home for lunch. All three were
from Sanaʿa and Muhammad al-Haimi was well known and respected by
the Bayt al-Maswari. In an expert manner they plastered the interior face
of the masonry wall lining the stairwell from bottom to top, leaving the
stonework of the central column exposed, and plastering the underside of
the winding staircase (see plate 48) in an organically sculptural way which
gave me the faint impression of being in the interior spaces of Antonio
Gaudi's Casa Mila in Barcelona.

All of the decorative brick relief elements were picked out in white (see
plate 49); the junctions between planes were highlighted; and the contours
of doors and apertures were framed with *guss* in typical Sanaʿani fashion,
rendering an almost gingerbread-like effect in unison with the other
buildings in the Bir al-ʿAzab quarter. Monitoring work on the exterior made
it clear that plastering the decorative elements calls for more than just

PLATE 49
Decorative brick relief picked
out in white plaster.

artistic finesse: the task demands a tremendous amount of nerve and, rightly, is reputed to be the most dangerous job in the building trade. "Look at them!", said one man behind me, tapping my shoulder. An audience of casual spectators had gathered with me on the street below. "Up there on that platform and playing about on that dome", he continued, "smoking cigarettes, chatting away, and paying no attention ... and no ropes or belts! No wonder we hear about them getting killed!". Others shook their heads in agreement. He was right. It did appear from their casual behaviour that the event was probably more unsettling for we, the uninitiated, watching from below.

Muhammad al-Haimi sat the whole while relaxed facing in toward the minaret on a small wooden platform that had two extended arms abutting the wall surface, and that was suspended by ropes which were secured inside the dome through the star-like perforations (see plate 50). His nephews prepared buckets full of *guss* on the calling platform at the top of the stairs, and passed these down to their uncle by rope. Muhammad could

PLATE 50
Muhammad al-Haimi
suspended on a platform and
plastering the exterior surface
of the minaret with *guss*.

move himself laterally a short distance in either direction by pushing off the wall in front of him with his dangling legs. To change his vertical position, he was required to get off the platform and lower himself by rope to the next level below his present work station, which landed him either on the balcony or on top of the square stone base. The position of the platform was readjusted by the nephews, one of which darted in and out of the narrow aperture high on the south side of the dome like a weaver bird from its nest, and scaled along the dome's surface using the perforations with the agility of a spider in its web, all in order to correct the position of the ropes.

When finally completed later that Summer, the 'Addil minaret was highly praised by local residents and by Sana'anis of other quarters alike. Many claimed that it was the most elegant minaret of Sana'a, and Muhammad al-Maswari, reticently proud, agreed that it was possibly his family's best work yet (see plates 51–52).

PLATE 51
The completed 'Addil minaret before its decorative brick elements were picked out in *guss*.

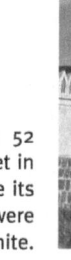

PLATE 52
The completed al-Ihsan minaret in Madinat al-Asbahi before its decorative brick elements were picked out in white.

201

II Master Builders: Intentionality and Expert Knowledge

The Brothers al-Maswari

Kull-an ʿala mahratu sati hakeem
Everyone in his profession is a wise *usta*.[5]

Usta, plural *asatiyyah*, referring to the Master of a trade such as building, is a colloquialism derived from *ustath*, the standard Arabic title for teacher or professor. This common title is used when addressing normally older men of higher rank who are considered to possess an expert knowledge or skill. *ʿAmm*, a less formal title also conferring respect, literally translates to 'uncle', and is more popularly used when speaking to those elders with whom one has more familiar, but not necessarily blood, relationships.

Many young labourers working in the service of the Bayt al-Maswari as well as for other *asatiyyah* in Sanaʿa customarily prefixed the first name of their elder superiors with *ʿamm*, for example, *ʿamm Muhammad* or *ʿamm Yahya* (the senior stone mason). However, it was less common to employ titles when addressing the younger Master Builders of the essentially gerontocratic order like Muhammad's son Majid or the young Dhamari *usta*, ʿAbdullah ʿAli as-Samawi. Their more limited experience was reflected by their respective positions in the hierarchy, and even when granted varying degrees of autonomy over individual construction projects, Majid and ʿAbdullah remained fully accountable to their senior masters. These newer members of the senior rank were not significantly older than many of the labourers and in several cases were younger. Largely by virtue of their comparative youth, they were not perceived to merit the same protocol as the two elder al-Maswari brothers, and their relations with the apprentice(s) and labourers was noticeably more casual (see plate 53).

ʿAbdullah shared living quarters in Sanaʿa with the migrant team of labourers from his village, but, when off duty, he actively asserted his superior professional status by means of the few belongings he maintained and through his discerning social practices. ʿAbdullah was consistently meticulous about his attire and was the only lodger in the shared residence to possess a metal spring cot. During tea breaks and lunch, he aligned himself closely with the al-Maswari brothers, and remained aloof when in the company of the lower ranked workers. He performed his tasks diligently, setting an example for his fellow villagers who had come, like himself, to the big city, and he preserved a somewhat serious, almost arrogant air. Unlike Majid, ʿAbdullah would not inherit the security of his position through blood and was thus compelled to 'earn' the respect accorded to him. By contrast, Majid had gained the reputation of being lazy and unreliable amongst his fellow workers, and his disposition was said by some to be as fickle as his work ethic. "He spends more time dragging on

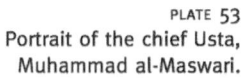

PLATE 53
Portrait of the chief Usta,
Muhammad al-Maswari.

his cigarette and gazing out over the rooftops of Sana'a than he does laying bricks", complained one labourer, "And then he'll holler at us if things fall behind schedule!".

Majid, a nickname derived from his real name Najib, had in fact pursued a career in building like his father and uncles largely because, as he told me, he thought it an easier option than his studies. He left primary school in the sixth grade, at the age of twelve he remembers, and began working as a labourer in the family business. Gradually he was placed under the direction of his father, Muhammad, and trained as an apprentice, performing the same tasks that Saat and his cousin Ibrahim were currently performing during my fieldwork. His long training was patterned after his father's own instruction under the supervision of his father, 'Ali Said, and only upon reaching his mid-twenties was Majid delegated the responsibilities associated with a Master Builder, being permitted to lay bricks and play a more active role in the design and decision-making processes. Having proved himself an adept craftsman, he was eventually assigned commis-

203

sions over which he had nearly complete autonomy, such as the erection of the minarets for the Dar Salam and the al-Sadiq Yahya Mesa'eek Mosques, but his work would nevertheless continue to bear the certification of his father's approval as head of the family business. Now in his mid-thirties and accomplished in his trade, Majid was, however, increasingly prone to taking lengthy breaks from the work site, protesting that he was *ta'ban* (tired) and in need of a rest. As Muhammad's only heir engaged in the family business, he could afford a certain insouciant attitude couched in the security of knowing that he could return when he wished.

Ta'ban was an adjective wielded oppositionally by members of the Bayt al-Maswari in depicting others of the same House. This form of slanderous commentary hinted at the tensions which lurked below the surface of relations between several family members, and which presently governed the division of labour between the four sons of 'Ali Said, the founding family member in the building trade. Initially, the existing antagonisms between siblings were not immediately apparent and were carefully concealed behind public formalities and a resistance to speak openly about family conflict with members outside the Bayt. It was largely through the web of gossip spun by the labourers that I came to realise the complexities in the relationships between the Brothers al-Maswari. *Ta'ban*, I discovered, might suggest more profound connotations than simply 'being weary': it was also a way of saying that something was 'washed-up' or 'finished', for example as in the phrase *"hatha Tareeq ta'ban"*, "this road is finished", referring to a state of disrepair. When referring to a person, it might suggest that the subject of discussion is also 'washed-up', possibly a heavy drinker, or someone with questionable and corrupted morals.

'Ali Said, who had made his apprenticeship in Sana'a under the tutelage of *Usta* Aziz Mayyad and *Usta* Muhammad Moqbal Mayyad, had trained all four sons born of his first wife in his adopted trade. All four young men continued to work together under his supervision, building residences and mosques, and later minarets for which they received their first commission around 1980. However, in 1996, just ten years after his death, it seemed that only two of the sons, Muhammad and Ahmad, were still fully engaged in the profession, and a third, Husayn, accepted only occasional commissions. When I first arrived in Sana'a, Husayn, the second eldest after Muhammad, was the *usta* in charge of laying new *habash* paving stones in the courtyard, or *hawsh*, preceding the entrance to the Hanzal Mosque in the Bir al-'Azab quarter.[6] Interestingly, both the Hanzal Mosque and the 'Addil Mosque for which Muhammad and Ahmad were currently building a minaret, were brought to life as characters in al-Khafanji's mid-18th century poem on the *Mosques of Sana'a*.[7] It was quite by chance that I found him working there during the brief period when I was residing up the road from that Mosque at the American Institute of Yemeni Studies, and although I had spoken with him I did not know he was an al-Maswari.

Muhammad and Ahmad had made no mention to me of their brothers, and it was not until I discussed the work I had seen at the Hanzal Mosque did they acknowledged Husayn as their sibling. After he had completed the paving project he took on no further work to my knowledge, and when I would occasionally inquire about him I was simply told that Husayn wasn't working anymore, "He's at home. He's *ta'ban*". From what I could determine Muhammad and Ahmad rarely fraternised with him, and were uncomfortable discussing him. I never learned anything further about Husayn, nor did I ever see him again.

While it's possible that Husayn may have been *ta'ban* in the conventional sense of being truly 'tired' (I have no evidence to suggest otherwise), the fourth brother, 'Ali, accentuated all additional connotations of that descriptive adjective. 'Ali was the youngest of the four, some seventeen years younger than his eldest brother Muhammad, and only one year older than his nephew Majid. He proclaimed himself to be an *usta* like his brothers, although it was confirmed that he had not been involved in a building project for some time. In fact there was something rather decrepit about 'Ali. His diminished stature was frail, and his face gaunt with dark withdrawn eyes that darted about as he rambled on nervously in conversation. He occasionally called at the building site of the 'Addil Mosque when walking about, and would climb the staircase to chat with his brother Ahmad at the top. Muhammad, after extending a noticeably curt greeting, mostly ignored the presence of his youngest brother during his visits. Initially I was eager to meet other family members and, oblivious to the seriousness of Muhammad's irritation, mistakenly invited 'Ali to join the rest of the builders to chew *qat* one Friday afternoon at my house in the old city.

'Ali was the first to arrive, and more than an hour early. I was still busy assembling trays of glasses, coolers of mineral water, and had not yet bought any "Canada" from the *baqqalah* (convenience store) next door. I made sure he was comfortable with his bundle of *qat* in the large *mafraj* (reception room) on the top floor and continued with my preparations. He was very uneasy and spent most of the time talking on the telephone, apparently to his elderly mother. Finally the rest of the builders arrived at the more normal time of two o'clock, and among the group was Ahmad, but Muhammad and Majid were both absent. I was told that they had to attend a *qat* session with clients in order to discuss future building projects, and that they sent their apologies. I found this mildly peculiar as both had confirmed with me the day before that they were coming. The following week I was surprised when told by 'Abdullah, one of Muhammad's sons who did not build but had been present, that his father and brother had refused to attend when they discovered that 'Ali had been invited: "'Ali is not a good man. He talks badly, doesn't pray, and he doesn't work".

The qualifications initially stated for 'Ali's not being a "good man" registered as being incompatible with what I had observed so far on the building site. With the exception of Muhammad who remained an utmost gentleman, the other workers including Ahmad and Majid certainly talked "badly" from time to time, and most of them rarely, if ever, prayed, but these qualities did not make them repugnantly bad. Likewise, Majid's younger brother, 'Abdullah, with whom I became good friends, didn't work in any capacity throughout my stay in Yemen, but he had no hesitations about casting that stone upon his uncle. There was evidently something more reprehensible about 'Ali's behaviour that was not being shared with me. Nearly two weeks after my qat session, several labourers came to me in the darkness of the winding staircase where I was working, and huddled around on the treads above and below to share their hushed insights on 'Ali. They had obviously discussed the matter and decided that I should be advised on the affair. "Don't let 'Ali in your house again", began one. "Why?" I asked naively, and a bit astonished by such a pronouncement. "He's a thief", responded a second, "And a liar" added another. "He also drinks too much, and Muhammad al-Maswari can't stand him for it", said the first in a horse whisper. "If he comes to your house he's likely to steal things. We don't trust him around the pockets of our own coats!" The pieces were coming together.

Some weeks later 'Ali did show up at my house a second time. It was late in the evening and he was obviously heavily inebriated. He knocked loudly at the door shouting out "Abu Bakir", as I was called by my Yemeni friends who found "Trevor" to be a strange and nearly impossible name to remember, and even more difficult to pronounce. I peered out a third storey window to tell him to lower his voice and that I was on my way down. Very pathetically he had come by to ask for money on his tottery journey home, and reluctantly I gave him a hundred riyals to see him on his way. He never came back to my house, nor did I ever see him at the minaret site again, leading me to suspect that he had later received firm warnings from his eldest brother. 'Ali was not tolerated because 'Ali was excessive. The stealing, if true, was certainly a condemned practice in Yemeni society, as it is in most societies, but drinking alcohol poses a more ambiguous question of 'right or wrong'.

With regard to the Iranian context, Tapper remarks that "the morality and legitimacy of wine drinking have been a constant subject for religious debate". Tradition records that "'Ali, the most prominent Shi'ite imam, said that if a drop of wine were to fall in a well, and if then a sheep were to drink from the well and rejoin the flock, he would not eat the meat of any sheep of that flock". The primary concern more generally, however, is to maintain 'aql (reason), and avoid intoxication that might release passion and emotions.[8] Substances which obscure 'aql have long been an issue of debate in the Muslim world, beginning with wine (khamr) – probably originally date wine, and later extended to include any alcoholic beverage – and later including "the nature and proper attitudes to coffee, tea, tobacco, *qat*, and harder drugs".[9] Conflicting attitudes toward wine tend to emerge within the Qur'an itself. Compare, for example, the following verses in *suras* 5 and 47:

> O believers, this wine and gambling, these idols, and these arrows you use for divination, are all acts of Satan; so keep away from them.[10]
>
> *(sura* 5: verse 90)

> The semblance of Paradise promised the pious and devout (is that of a garden) with streams of water that will not go rank, and rivers of milk whose taste will not undergo a change, and rivers of wine delectable to drinkers, and streams of purified honey, and fruits of every kind in them, and forgiveness of their Lord.[11]
>
> *(sura* 47: verse 15)

Although drinking *khamr* is generally conceived as erroneous, and the prohibition against it made in the name of the Holy Qur'an, a significant enough minority of Yemenis indulged in the occasional swill of Johnny Walker that, for many, the moral question of consuming alcohol had become one of gradated degrees as opposed to simply either 'right *or* wrong'. I was told that other vices, for instance male engagement in adulterous or homosexual affairs, might be conceived in the same manner. All the Yemeni Muslims with whom I socially engaged (with several exceptions) publicly professed that drinking is *haram* (forbidden), but in fact many judged individual cases in a less exacting manner. Rather, what was perceived by a social group to be 'excessive indulgence' with respect to certain types of vices, and not the mere act of indulging itself, frequently served as the dividing line between acceptability and repugnance. Certainly, many other builders aside from 'Ali had consumed alcohol but managed to resist condemnation from their colleagues by staying within the socially acceptable bounds negotiated and defined by their peers. Not surprisingly, most individuals entertained two levels of moral judgment: that which they interpreted as being decreed by the Qur'an and inculcated in the process of their Muslim socialisation;[12] and the more ambiguous degrees of tolerance which were informed by the former, but continually negotiated, defined and redefined within the social context.

Muhammad, as the head of both the family and building operations, made a concerted effort to dissociate himself from his youngest brother. "'Ali", Majid pointed out, "could cause building patrons and the public to look unfavourably upon the Bayt al-Maswari". Interestingly, Bonnenfant reports that "Parmi les bons *usta* actuel de maçonnerie, citons la famille al-Miswari. Trois fils de feu l'*usta* 'Ali Sa'id al-Miswari, l'aîné Muhammad, puis Hasan et Ahmad, ont edifié dans les années 1980 plusieurs nouveaux minarets de Sanaa hors les murs ... ",[13] and makes no mention of the fourth brother, 'Ali. There was a conscious recognition, or belief, by all working members of the Bayt that their affiliation with the construction of religious edifices coloured the public's expectations of their conduct. The intentionality driving Muhammad's decisions and performances as both a builder and a social being was largely formulated in response to these perceived expectations, and correspondingly he shouldered most of the responsibility in producing and preserving the reputation of his building House. I will address the topic of intentionality in greater detail later in this chapter.

Reputation constituted a very significant component in the formation of the team's identity as Master craftsmen, and all played a role in bolstering the magnitude of that reputation not only through the performance of their tasks, but equally by means of the often exaggerated reports they made. Ahmad announced in my presence at a *qat* chew that "the Bayt al-Maswari is the only notable House of traditional builders still working in Sana'a. Most builders receive a commission here and there, but only we have a continuous run of prestigious projects". It must be considered that the al-Maswaris were not necessarily familiar with building projects in the walled city since they neither lived nor worked there, but there was nevertheless a great deal of truth in Ahmad's boastful claims. At the time of my study, several of the Master Builders at various building sites around Sana'a, building with masonry in a traditional Sana'ani style, were not in fact from the capital. Also, during the year that I worked for the al-Maswaris, they erected two major Sana'ani mosque minarets, a smaller minaret in a nearby village, and the brickwork for two new houses in Sanhan belonging to General Mahdi Maqwallah, a brother of the country's president. At the time of my leaving, Muhammad and Ahmad were in the process of negotiating the commissions for four new mosque minarets including one of seventy meters for the al-Mutawakkil Mosque adjacent the Maydan Tahrir. Unquestionably they had cornered the 'minaret market'. The latter proposal would easily be the highest minaret in Yemen, but construction had not yet commenced when I returned to Sana'a for a brief visit in the Winter of 1998.

Ahmad told me confidently that the family legacy in the trade would be continued by successive generations, partially directing his prophecy to the ears of his son Ibrahim and his nephew Osam sitting nearby. Both boys squirmed and forced nervous smiles. I noticed for the first time that one of Ahmad's eyes was slightly opaque and he was probably developing a cataract. "I will teach my son", he continued, "and he will teach his". There was something undeniably noble about his aspirations to keep the wealth of trade knowledge and the professional status in his family, but with continued observation, Ahmad's goals seemed increasingly improbable. The assurance of the younger generation of al-Maswaris, including Majid, in knowing that their place was more or less secured in the trade led them to be lazier, more impatient for rewards and recognition, and often disrespectful, in comparison to colleagues such as ʿAbdullah al-Samawi, Saat, or Muhammad the Tobacco Chewer, none of whom had blood ties with the family. Majid, with his father's political lucidity and a sharper, cutting wit, regularly antagonised his Uncle Ahmad who responded with garbled enraged outbursts between *qat*-green teeth and sputtering leaves. This type of aggravation would not be tolerated from any other junior builder. Likewise, Ibrahim, as discussed in the last chapter, was unruly and showed little promise in becoming qualified for taking on the responsibilities of an *usta* and providing an example for his labourers. More so than Majid, Ibrahim's disposition on site betrayed his belief that respect was something he *deserved* rather than something he had to *earn*.

Finally, it must be contemplated whether a professional sense of discipline, considered in the second chapter as a key component in the formation of a Master craftsman, can be efficiently and economically instilled in successive members of the same family through an apprenticeship system. Coy writes that it is not uncommon for Masters in various trades to resist training their own children because of the inherent problems in disciplining one's own kin, and to "avoid the disruptive combination of the resultant role set, father/master and apprentice/son."[14] The sceptical nature of my observations regarding the potential family heirs to the head of the building practice are therefore not surprising when considered in this context. In contrast to Ahmad's optimistic prognosis, the common Sanaʿani assertion that 'the great Building Houses never endure longer than three or four generations' rings with legitimate threats to the survival of the Bayt al-Maswari. Having recognised this predicament to some extent, in addition to acknowledging the migration of young al-Maswari males to alternative professions, Muhammad has chosen to propogate his skills via non-related males. While striving to preserve the knowledge of his trade he has simultaneously engaged in destroying the very power base by which he and his family once controlled the expert discourse.

Earning an Expert's Expertise

ʿAbdullah as-Samawi returned to work with the rest of the team on the minaret of the al-Ihsan Mosque after completing the commission for a small five meter minaret in a village near Sanaʿa. Muhammad felt confident in assigning ʿAbdullah to supervise such projects since, by this time, he had been working for the Bayt al-Maswari for over ten years and had proved himself a diligent and skilful worker in the eyes of his Masters (see plate 54). The stone base of the al-Ihsan minaret had already been well advanced under the direction of the two elder brothers assisted by their *thana'a*, Gayyid, and when ʿAbdullah arrived he was

209

PLATE 54 'Abdullah as-Samawi and Ahmad al-Maswari setting
courses of bricks at the top of the walls of the 'Addil minaret.

instructed to assist Yahya Faraj, the *muwaqqis*, in preparing the blocks
of dressed stone. There was no immediate need to recruit the young *usta*
to the top since the stone courses were being set by the two senior
Masters at a satisfactory rate. More importantly, Muhammad maintained
that the practical experience with Yahya would be beneficial for
'Abdullah and would enable him to hone his masonry skills. Later,
once work had commenced on the brick portion of the tower, he was
appointed once again to his familiar position as builder.

Of the two senior brothers, Muhammad in particular kept a close eye on
the younger *asatiyyah*, as well as on the rest of the work team, guiding their
practices and correcting their mistakes when necessary. Through years of
experience he had acquired an astonishing ability for effecting his own tasks
while simultaneously monitoring the progress of his fellow workers. Once,
I watched him turn quickly in the direction of 'Abdullah while in the
process of setting his own brickwork on the exterior face of the tower, and
inform the young builder that the course-work he was laying around the
wall of the stair shaft was not conforming to the prescribed radius.
Accustomed to this form of critical commentary, 'Abdullah did not raise
issue with his Master's seemingly off-handed judgement, but instead, with
furrowed brow, verified the curvature of his completed work with the radial
rope attached to the axial metal post. Muhammad was correct. The section
of circumferential course-work to which he had referred was indeed leaning

inward by little more than a slim centimetre and would require adjustment. Perfect alignment was not only an aesthetic hallmark of the al-Maswari craftsmanship, but an absolutely necessary quality if these towering edifices were to preserve their structural integrity.

In nearly all cases, a young builder remained under the tutelage of his seniors even after his responsibilities and wages had been augmented and his status as an *usta* had been confirmed by his Master and recognised by his peers. This was true for all the building teams with whom I had either worked or called upon including that of Usta Husayn 'Ali Humdi al-Turaygi, an expert in the construction of public bathhouses, or Usta Naji al-Khowbani placed in charge of building a new school. Both these Masters, like Muhammad and Ahmad al-Maswari, employed labour teams which included several permanent junior *asatiyyah*. In every practical sense, the teacher/student hierarchy of the apprenticeship persisted until the builder became the Master of his own Building House. There were numerous important skills to be acquired in supplement to the disciplined performance of trade related tasks. These included the establishment and nurturing of client relations; the estimation of building costs and materials; the management of physical and personal resources; and most importantly, the seemingly intuitive knowledge about all phases of the design and construction process, from the micro to the macro level, from the practical to the conceptual level. All of these, and the latter in particular, required sufficient time and immersion in every dimension of the practice, and ultimately the transgression of *being* from a level of simply 'understanding' to one of an 'intentionality' steeped in a *directedness toward* competent professional practice.

From the observations I made on site, it seemed evident that 'Abdullah, in comparison with Majid, was better suited as a candidate for one day becoming an accomplished head usta. Likewise, as discussed in the previous chapter, Saat had proven himself a more responsible and attentive assistant than Ahmad's son Ibrahim, though blood might complicate the decisions for promotion. Both 'Abdullah and Saat had submitted themselves entirely to the disciplinary measures of their training from the time they began work as labourers, rendering themselves subservient to their Masters, and progressively charting their career paths in a direction that would lead to the mastery of their trade. Neither of the two were in a position to directly contest the authority of their superiors and both had to juggle their abilities to impress with their abilities to 'steal' knowledge by careful surveillance, all the while safeguarding their precariously held posts against would-be usurpers. What they possessed in terms of status, skills, and knowledge had been earned through an adherence to the rules of the system and an advantageous manipulation of the sentiment and decisions of those who produced it.

Majid, on the other hand, found himself in the family profession primarily because it was presented as an easy career choice, and, as he told me, he had not shown a strong aptitude for formal schooling as a child. Though he was a skilled craftsman, the training he received from his father and uncle, and the assurance of a livelihood in the trade, rendered him a less self-directed and motivated professional than his counterpart, 'Abdullah. On one occasion, toward the completion of the 'Addil minaret, he left his work post and invited me to follow him to the top of the minaret. We climbed out through an opening at the top of the incomplete dome and sat there together, enjoying the magnificent view, the curved vaults of the dome disappearing from view below us and giving the thrilling sensation that we had been suspended in mid air high above the city. Majid took breaks frequently and enjoyed them better with company. We chatted about the state of the current project and about those to come, about the progress of my study, and about Majid's career as a builder. He confided that building was a hard profession that leaves one's body tired and aching. "My dream is to be given one hundred and fifty million American dollars, and no less" he told me dreamily, "and then I could quit building and travel the world, to Egypt and India, and I'd return with four wives. Only one would be a Yemeni woman and the other three I would take from Europe. When ultimately satisfied, I'd make my pilgrimage to Mecca and then I would be ready to die a happy and fulfilled man". Before we left our perch he gazed down below to the roof of the mosque where another team of labourers were plastering the roof with *qaddad*,[15] and sighed, "At least my work is more pleasing than that!".

After completing the brickwork on the General's new house in Sanhan, Majid did not join the others at the al-Ihsan minaret in Madinat al-Asbahi, but chose instead to take a lengthy break. He remained for the most part with his wife and children at his home which was connected to his father's in the Bir al-'Azab quarter, and from my knowledge he never visited the new minaret project during its construction. At the time I left Yemen, Majid had been off work for several consecutive months.

The Design of Expert Knowledge

Occasionally in the evenings after chewing *qat*, 'Abdullah as-Samawi took pen to paper and drew brick patterns and produced rendered drawings of minaret elevations. Ahmad al-Maswari also sketched out decorative brick patterns on scraps of paper in order to calculate geometries, but only 'Abdullah approached the endeavour with a somewhat artistic demeanour. Twice he proudly presented me with elevation drawings of hypothetical minarets which he had executed with a straight-edge ruler, one rendered in red and blue ball point pen on the backside of a torn poster, and the other in lime-green, pink, and black coloured pencil (see figures 17 & 18). The decorative patterns shown in the ink drawings are reminiscent of those

FIGURE 17
Ballpoint pen drawing of two minarets by Usta
ʿAbdullah as-Samawi.

FIGURE 18
Colour pencil drawing of a minaret
by Usta ʿAbdullah as-Samawi.

which were carried out on the 'Addil minaret, and the patterns in the pencil rendering resemble the Zuheiry Mosque minaret built near the western edge of the city. His drawings seemed awkwardly proportioned to my eye, exhibiting over-sized towers supported on stunted black stone bases. The prayer calling platforms of the two minarets in the ink drawing are incommensurate with the dimensions of the towers. In the coloured pencil drawing, the Christian calendar date of "1995" inscribed in Arabic numerals above the door is curiously written backwards. Nevertheless, 'Abdullah's drawings effectively captured the essence of the 'Sana'ani-style' minarets which he had built with the Bayt al-Maswari, and, interestingly, they are the only complete renditions of these structure produced by the builders with whom I worked. In fact, the other builders demonstrated little, or no, interest in 'Abdullah's drawings.

With the start of the decorative brickwork on the al-Ihsan minaret, I began to inquire about the production of the design. 'Abdullah ambitiously staked out his claim to the authorship of this work, boasting that the geometric solutions for the intertwining diamonds and chevrons came to him while he was chewing *qat*. "The inspiration was so clear that I didn't have to draw it!" he told me. Then, with a fixed gaze he pressed his forefingers to his temples and, lowering his voice, he continued "It's all here in my head". I asked him what it would look like when completed, and he paused momentarily in thought, then crouched down to pick up a small sharp stone and drew a rough version on the dirt ground between us. Truthfully, it was difficult at the time for me to discern his pattern from any other Sana'ani brickwork I presently knew. Oddly, Muhammad al-Maswari, who had been present, later claimed that it was in fact himself who engineered the scheme. I approached 'Abdullah with this counter-claim to which he responded unconcernedly that he had indeed thought up the initial idea and his Master had accepted it. Muhammad's stamp of approval effectively transferred the authorship into his own hands, and he likely embellished the design in conformity with the repertoire of his past work. I couldn't help but to think of the countless painters who had apprenticed in the studios of Peter Paul Rubens, filling his canvasses with fleshy forms and drapery, yet never experiencing the glory of seeing their own signatures in the bottom corner.

One morning I brought a postcard from Oxford to the site in order to show it to the builders. It was a photograph displaying the 15th century fan-vaulted ceiling of the Bodleian Library with its carved stone bosses at each intersection of the ribs (see plate 55). I was curious as to whether these men would have any interest or understanding of the late Gothic masonry structure. A group gathered tightly around me, as was normally the case when I brought photos or something 'exotic' along to work, and some of the labourers tugged at the edges of the postcard in order to get a better view. I explained that it was a photograph of a ceiling constructed in stone,

PLATE 55 Postcard of the stone vaulted ceiling of the Bodleian
Library, Oxford.

but the abstractness of the photo sustained most attention spans for only a
few moments, and I was soon left with Ahmad and Majid. At this point I
was able to safely hand it to Ahmad, and he studied it for a few moments,
rotating the image completely, then exclaimed with mild amusement
"Stone! hmmm", before passing it on to Majid and heading into the
staircase of the minaret. Majid showed the most interest, studying it
carefully in an attempt to understand what the expression of the stone
could tell him about the structure, and he asked several questions about the
assembly. "I'll build a ceiling like this here in Sanaʻa!" he said laughing, and
handed the card back to me.

In the same manner that I used the postcard and other images to convey
ideas about structure and space, and to invite a discourse on these issues,
some of the builders, and most frequently Ahmad, produced thumbnail
sketches for the decorative brickwork designs in order to objectify their
ideas and elicit a consensus among themselves. One morning, at the start of
the shift, Ahmad pulled out a neatly folded piece of loose-leaf paper from
his coat pocket and opened it to reveal simple line drawings of brick
patterns that he had sketched the night before. They were crudely executed
in ball point pen, but served to sufficiently convey his ideas and work out
the assemblies of the geometric patterns. As always, he avowed that
chewing *qat* had uncluttered his mind and provided him with the necessary
stimulus for 'envisioning' the design. "*Qat*" was by far the most recurrent
explanation I received from Yemeni craftsmen, including carpenters and

decorative plaster workers, when I inquired about their creative processes. In his studies on *qat*, Varisco states that "poets actively seek inspiration while chewing",[16] and that labourers he spoke with claimed that *qat* increases their energy levels.[17] University students that I spoke with also alleged that *qat* chewing ameliorated their writing and study skills. *Qat*, being an amphetamine-like drug, produces similar effects as a strong cup of coffee, and when chewed, the drug both relieves fatigue and hunger, and increases alertness and concentration. I was repeatedly told that "*Qat* is *our* whisky. You drink, ... we chew."

On his pieces of paper, Ahmad had produced two drawings side by side. The first was a single line sketch of the geometries of interlocking diamonds and chevrons, referred to as *haras* and meaning 'to guard or protect'. In the second, he had conveniently modularised his design with the use of the lines on the standard loose-leaf page, whereby the space between two lines represented the thickness of an *ajurr* brick. The first drawing represented his conceptual design of the overall form, and the 'making' of the second served as a means to solve the practicalities of the assembly with drawn representations of brick units. Once the form of the design had been resolved in the first sketch, its shape having been conceptualised in accordance with the known array of available carved bricks,[18] it could then be translated into the 'working drawing'. From this second drawing, the exact quantity and type of carved bricks required could be specified. Since Sana'ani architectural decoration is nearly always composed of the sequential repetition of a stylised pattern, it was necessary to draw only one module of the decoration completely. Once the number of vertical brick courses required to complete the individual design module had been determined, its metric height was known, and therefore the number of repetitions requisite for achieving a specified vertical height of the tower could also be pre-determined. The conceptual *making* of the edifice thus began with the concept of the brick unit, and the successive conceptual manipulation of the spatial relations between units to *make* larger units, or *objects of knowledge*.

All of the designs produced were variations on a common theme of rectilinear motifs that, through repeated use over centuries, had come to represent the dominant form of Sana'ani architectural relief decoration. Their production, within an aesthetic framework which tolerated limited variance, was less an expression of emotion than it was a "science", and the craftsman, like the Muslim artist described by Titus Burckhardt in his foreword to *Geometric Concepts in Islamic Art*, "willingly subordinate(d) his individuality to the, as it were, objective and impersonal beauty of his work".[19] Prior to the revolution in the 1960s and the subsequent expansion of the city beyond its walls, these decorative forms constituted a dominant visual element unifying the urban landscape of Sana'a, and serving that architectural and metaphysical unity which Nasr claims "always dominated over the parts" of what he identifies as the 'traditional Islamic city', but which allowed for "the growth of the parts, but always in relation to and in harmony with the whole".[20]

The dominant geometric theme of the Sana'ani brick decoration has been constrained firstly by the historically consistent fabrication and use of square, flat, *ajurr* bricks, the dimensions of which have varied only marginally; secondly, by the limited repertoire of 'carved' units that are produced by the builders (the repertoire itself is largely dictated by the form and quality of the available brick and the manual tools used for carving); and thirdly, by the finite range of possibilities for assembling these carved units in decorative arrangements. However, though finite, the exercise of documenting and cataloguing all of the existing motifs and their variations would present a formidable task for any researcher,[21] and the individual motifs tend not to be named or ascribed with specific symbolic meaning by the builders. As noted by Murata and Chittick in *The Vision of Islam*, the abstract nature of Islamic art represents ideas rather than things, and its use of abstract and geometric forms serves to remind people of "the divine beauty by detaching that beauty from this world; that is, from the things that figurative art attempts to represent".[22] In addition to the intrinsic physical constraints of the brick material and tools, and the Islamic bias toward abstract and geometric representations, the *conceptualisation* of these patterns as 'traditional', or *taqlidi*, and the socially constructed values currently associated with traditional architecture, upheld by both client and builder, should perhaps be recognised as the most significant factors constraining the development of radically 'new' decorative motifs.

It should be noted that in previous decades, new money and external influences created client demands which threatened to eradicate indigenous design and construction technology, and replace it with an urban building stock steeped in Western Modernism. However, political and economic upsets weakened the potential for a modernist renewal of Sana'a in the style of Saudi Arabia or the Gulf countries. More positively, an expanding international interest and burgeoning enlightenment of local public appreciation for Yemeni heritage have also served to shift that trend. This shift has propagated a variety of aesthetic courses, including a revival of 'traditional' building, and also the flourishing of an original contemporary architectural style evinced in the expanding, and usually wealthier, suburbs of the capital. The erection of reinforced concrete-frame apartment and office towers also persists along several of the principal commercial arteries of the city, but the architects and contractors of these recent works have visibly made more concerted, though often questionable, efforts to incorporate local materials and craftsmanship. These often amount to little more than the appliqué of multi-coloured stone on the facades and the banal inclusion of carved plaster fanlights of stained glass, or *gamariyyah* windows, but there has not, as yet, been a return to the steel and glass construction like that of the infamous Yemeni Airline office tower built in the late 1970s.

Builders like the al-Maswaris play an active role in the often combative discourse between traditional and modernist architecture. Late one morning, resting on the top of the walls of the 'Addil minaret with the two elder Masters, I asked them for their impressions of the new office towers being built in the city. Ahmad shrugged his eyebrows and dismissed them as "no good", and proceeded to sip his tea from a tin can. Mohammed turned his gaze toward the half a dozen or so concrete structures rising from Zubeiry Street and looming on the horizon to the south-west of us. He stared fixedly in their direction, remaining silent for a few moments before responding that "This (the minaret) will be here for a long time after they've (the concrete towers) crumbled and disappeared. Our construction is *qaweey jidan* (very strong) ... we use brick and stone, and we don't need architects and their plans."

One popular tale told by traditionalists tells the story of a Master Builder working in the service of the ruling Imam earlier this century. After building a palace for the Imam in a wadi outside the city, he was ordered to construct a mosque nearby where the ruler could pray. The Imam told his builder, who never relied on the use of surveying equipment, that if he could perfectly align the prayer niche, or *mihrab*, with Mecca, he would be rewarded with grains, but if he was mistaken in his calculation he would be punished and thrown into prison. After the construction was completed, the Imam summoned an engineer to the site to verify the orientation of his new mosque with compasses and modern survey equipment. The alignment with the Holy City was exact and the Master Builder reaped his reward. The story clearly evokes the powers popularly thought to be wielded by the traditional Master Builder by emphasising an apparently specific event rather than proposing theoretical inferences about the content or nature of the builders knowledge.[23] It reinforces their independence from, and superiority over, technical assistance since all knowledge is "in their heads". The telling of such tales can be recognised as an integral part of the expert discourse, serving to produce and reproduce their power over the trade knowledge in the eyes of the public.

'Ilm, or knowledge, is the most frequently used word in the Qur'an after the name of God,[24] and accordingly has become associated with the 'sacred', and necessitates safeguarding against questioning forces that may threaten its credibility. Eickelman remarks that the religious sciences constitute "the totality of knowledge and technique necessary in principle for a Muslim to lead the fullest possible religious life [and is] the most culturally valued knowledge".[25] 'Ilm includes such disciplines as Arabic grammar, principles of jurisprudence and astronomy,[26] and was also used by Yemenis to describe scientific, technical or specialised trade knowledge. More importantly, in the case of traditional knowledge, what must be safeguarded is not so much any actual 'knowledge', but rather, what Pascal Boyer identifies as the causal connection between "the state of affairs described and the actual person making the utterance".[27] The utterance is judged by the public to have a truth value if such a causal connection can be drawn between the utterance as an event and the 'customised' person who produces it.[28] The utterance as an 'event', and less importantly as a 'transmission of information', renders it more defensible against questioning and challenge: in the context of traditional discourse, the illocution, and not the content, becomes the most salient factor in modifying the public's conceptual representations of the 'order of things'. Therefore, in the building context of Sanaʿa, maintaining power over trade knowledge can be interpreted as producing and reproducing power over the 'causal connection' that links the traditional builder as 'customised person' (and his associated *maʿarifa*) with the 'customised utterances' he produces: importantly it is the society's perception of this 'causal connection' which imbues the Master Builder's 'traditional knowledge' with a truth value. In producing and reproducing such customised tales as the story of the Imam's Master Builder, the traditional builder is actively engaged in processes which reconstitute his own identity as a customised person wielding customised power, and thereby perpetuates the causal connection which links his two forms of 'knowledge', both *ʿilm* and *maʿarifa*, with veracity and credibility.

When Ahmad was asked how he and his family members acquired the knowledge and technical expertise necessary for erecting their first minaret for the Husayni Mosque in 1980, Ahmad boasted to me that they had never studied existing minarets, but simply relied on the trade expertise derived from their father, ʿAli Said. I was puzzled, and persisted in asking him how they were able to devise the correct proportioning of their first towers without the use of some measured precedent. He responded that his understanding of aesthetic proportion, like his structural abilities, was completely intuitive, and something honed through practice, not study. Ahmad asserted that "the proportions of the minaret tower are correct when the minaret fills my eye", simulating his magical metaphor by spreading the distance between the tip of his index finger and thumb in front of one eye. His vague pronouncements were typical of the builders' responses, and were more effectively aimed at structuring the audience's perception of his 'traditional' knowledge as a cultural fact to be accepted and revered, rather than illuminating the source and content of that expertise.

Nearly six months later, ʿAbdullah al-Maswari, a brother of Majid, informed me that site surveys of the Musa Mosque minaret conducted by his family had served to guide their early structural and design decisions. The Musa minaret was built in 1747 in a southern quarter of the walled city, and is considered by many as the ultimate paradigm of Sanaʿani minarets. I returned to Muhammad with this potentially controversial insight offered by his son, but he now openly acknowledged that they had indeed looked to the Musa and other old city minarets for inspiration. He qualified his concession, however, by insisting that no one had taught them 'how' to build these towers. Consistently the implication was that the know-how was an intrinsic part of their trade knowledge as Master Builders. This further supports my claims in the previous chapter that the type of expertise inculcated through the processes of 'making' during the apprenticeship produces an essentially non-objectified form of knowledge that is neither propositional nor amenable to scrutiny. In summary, the builder cannot explain 'how' he knows. Other researchers have expressed frustration over what they have conceived as a deliberately constructed wall of secrecy, without fully comprehending the limited potential for propositional engagement with this kind of 'knowing'. After interviewing six *asatiyyah* for his study on tradition versus modernity in Sana'a, al-Sabahi concludes that "despite all this effort, only one Master Builder supplied enough detailed information to be of use" in his research. He found that the others were "too busy to go into detail".[29]

In conclusion, the struggle to define the source and content of the *usta*'s expert trade knowledge confronts two formidable hurdles: the first, the incapacity of the traditional builder to explain 'how' he knows, has already been discussed at length in the third chapter and elaborated by further case studies in this section. The nature of the second hurdle reflects the notion of a 'deliberately constructed wall of secrecy', but this must be contextualised within a broader framework of 'intentionality'. Briefly, the reluctance to qualify sources and methods of knowledge, and the frequently misleading or embellished responses, should be considered as part of a 'customised discourse' which aims to produce and reproduce 'customised persons'. In the following section I will investigate the subject of human intentionality as the driving force behind the constitution and reconstitution of consciousness and identity, grounding the exploration in my studies of the Master Builders of the Bayt al-Maswari.

Expertise and Intentionality

Intentionality is not an ultimate explanation of the psychic but an initial approach toward overcoming the uncritical application of traditionally defined realities such as the psychic, consciousness, continuity of lived experience, reason.

Martin Heidegger[30]

In this section I will develop a model of intentionality grounded in a combined understanding of consciousness and experienced temporality, looking to Heidegger's theory of time-consciousness for direction. I will argue that intentionality drives all decision making processes at every level of our human existence and that this process represents the mechanism which effectively structures our sense of temporality, as well as the continual present-ness of our experienced identity. The discussion which follows has been divided into two parts: the first aims to anchor the argument in a specific philosophical discourse and develop a conceptual model of intentionality which will lend itself to further investigations into the Master Builder's processes of decision-making and identity construction; the objective of the second part is to introduce a level of individual agency to the model and demonstrate how levels of intentionality can be honed through practice, dedication, and experience.

Part 1: A Working Model of Human Intentionality

Franz Brentano, whose ideas immensely influenced the thinking of Husserl, and later Heidegger, determined that intentionality is what distinguishes the mental. Intentionality is the *sine qua non* of consciousness in Western phenomenology, inciting the experience of *distinctiveness* that is perceived between subject and object: consciousness is always *of something*.[31] Brentano submitted that only mental events exhibit the property of being directed towards an object, therefore to think is always to think about something, and likewise to perceive is always of something.[32] The *something*, or object of consciousness, toward which mental phenomena are intended is necessarily an entity that lies within the subject's awareness, and may equally include other mental phenomena.[33] Intentionality, therefore, can be understood as structuring all lived experience, or, in fact, consciousness, and must not be misapprehended as a secondary relationship that ensues between the subject and object.[34] Intentionality *is* the relation between the two. Heidegger writes:

> just as intentionality is not a subsequent co-ordination of at first unintentional lived experiences and objects but is rather a structure, so inherent in the basic constitution of the structure in each of its manifestations must always be found its own intentional toward-which, the intentum.[35]

Intentionality is being toward something, and is therefore anticipatory of possible 'futures' of present-ness. In this sense, the very structure constituting consciousness, or lived experience, is always ahead of itself in a temporal sense, or 'ecstatic'. This model of a temporally ecstatic consciousness for Heidegger is a human-acting-consciousness (or *Dasein*), and its ecstatic quality imbues it with the motivation necessary for human agency, or, in other words, the capacity to incite processes of action upon one's environment.[36] Likewise, it is this inherent intentional structure which is responsible for creating our conscious concept of time, or *time-consciousness*, and enables us to experience the temporal properties, albeit a (culturally and socially informed) construction, of our objects of consciousness.

In order to apprehend a theory of intentionality grounded in the principles of time-consciousness, we must grapple with the underlying notion of 'temporality' upon which it is based. Heidegger suggests that "The meaning of the Being of that being we call *Dasein* proves to be temporality" therefore necessitating "an original explication of Time as the horizon of the understanding of Being".[37] Human-acting-consciousness, as I have suggested here, is always ahead of itself, displaced existentially in the future. This future is not a predetermined temporal frame awaiting actualisation in a continuous stream of present-ness, but is constructed by the conceptual projection of alternatives intended-towards, which I will refer to by Husserl's terminology as 'protentions'. Equally, lived experience is a product of, and subject to, its 'history', construed as memories, recognised here to be in a dialectic with present experience. Therefore *dasein*, or the nature of intentionality, is constituted by the past from which it is projected or *thrown*. Heidegger has called this the *facticity of Dasein*, but I will borrow Husserl's terminology, and refer to this past-ness of human-acting-consciousness as 'retentions'. Finally, and ultimately, lived experience is perpetually actualised, or *ensnared*, in the present (*verfallen*). Consciousness of the stream of lived experience is firmly anchored in the present, and therefore the present is secured a temporal prominence.[38]

At any given moment, or perhaps it would be more accurate to state for any given experience (since temporality is a product of experience), all three temporal modes are in simultaneous operation. Consciousness is thrown from its past, anchored in the present, and intended-toward the future. Therefore present experience, or human-acting-consciousness, is actualised only through its unity with a conceptually constructed past and future. This overarching temporal structure, driven by the dynamics of intentionality, enables us to experience temporal extension and temporal passage in our objects of consciousness, and ultimately co-ordinates our multiple temporal perspectives. The passage of time, or for that matter, time itself, can therefore be understood as a product of consciousness. In further explaining Heideggerian notions of time passage, McInerney writes:

> the temporal extension or duration of human Being is not a matter of occupying contiguous points of time, but of being ecstatically 'outside of itself' and of 'stretching itself along'. All three ecstasies of temporalizing activity form an intrinsic unity and operate together in producing the passage of time.[39]

Past-present-future time is what Heidegger referred to as 'world time', and it is *in* this time that we encounter other entities in terms of experiencing. World time is, by nature of its construction, action-oriented in that every moment is conceived as being toward some potential moment or action. The capacity to maintain or reformulate the alternatives intended-towards demonstrates agency over the projection of our specific future, as does the actualisation of that future in the present demonstrate the power to exercise action upon our environment. "This active making present is the source of the passage of time".[40] Our conceptualisation of a passing stream of "nows" was formalised in Newton's theory of absolute time which continues to dominate our shared ideal of time. Nevertheless, Heidegger proposes that it is the universal access to the *now*, or coevalness, which makes time public, and enables us to *share* experience and a common, presently constructed, *historicity*. In Kantian convention, Heidegger maintains that the mechanics of consciousness open up a system of meanings, rooted either in the particular historicity of the individual or in that shared by a community of *daseins*, which, notably, pre-determines that which is intended-toward, and likewise, that which is encountered experientially.

Pursuing these developments, I suggest here that protentions, our human intended-towards, are necessarily constituted by retentions, or conceptual representations of the 'past'. It follows that retentions are the essential building blocks of conceptualised possible futures, or 'that towards which we intend' in action-oriented world time. Therefore the past is not merely 'behind us', but also 'in front of us', providing an overarching temporal frame which structures the meaning of our present experience. Intentionality can be understood as the driving force of consciousness, or in fact the *essence* of consciousness. In 1890, William James suggested that consciousness is more like the steam engine that powers the locomotive. However, in line with Heidegger, I would substitute James's 'consciousness' with 'intentionality' since the former is apprehended as a product of the latter.[41] Not only does intentionality sustain a perpetual mechanism of experiencing for the duration of one's existence, but it also structures the temporal dimension of experience necessary for both intellectual orientation (with priority on a conceptualised future) and the pragmatic survival of an individual (and species) in an action-oriented world of meaning. Flanagan advocates a similar Darwinian position, writing that "our capacity to think ahead, recognise novel situations as harbingers of good or ill, and to speedily and imaginatively solve problems are/were certainly key to the survival and proliferation of our species".[42] I would add here that our ability to *anticipate* is also key to the constitution and reconstitution of individual and group identity, like that of the 'traditional' Sanaʿani Master Builder (where the qualifying characteristic of 'traditional' is constituted and reconstituted by the builder(s) himself/themselves).

Before turning to the specific case of the builders, it may be beneficial to introduce relevant neuro-biological data to further support this working hypothesis of intentionality. It has been proposed by A.H. Klopf in *The Hedonistic Neuron* (1982) that neurons are in fact goal-seeking. Neurons are not only the material building blocks of the brain, but their innate goal-seeking quality can be traced to the level of Being, or consciousness:

> intelligent brain function can be understood in terms of nested hierarchies of heterostatic goal-seeking adaptive loops beginning at the level of the single neuron and extending upward to the level of the whole brain.[43]

Neurons, or nerve cells, which comprise the brain and nervous system, are made up of essentially the same parts as other cells, except they have the capacity to produce electrochemical pulses. The production of these pulses enables the neurons to communicate with one another and with other cells in the body. In their initial development, neurons synapse, or connect-up with, other neurons to form networks, and like other living organisms neurons share the inherent tendency to act upon their environments. Following Klopf's thesis, the synapsing process of neurons can be understood as 'goal-seeking', or propelled by intentionality. However, it should be noted that there is no proof that this goal-seeking is driven by a predisposition at the neuronal level to increase 'pleasure' and minimise 'pain' through a strategy of optimal firing (of electrochemical pulses), but rather the evidence points more to a highly predisposed genetic encoding.[44]

In the neocortex, the most recent evolutionary development of the human brain responsible for mediating higher order perception and cognition, neurons are organised into a series of feed-back loops carrying back immediate responses to the neuron's own activity within its environment. Laughlin, McManus, and d'Aquili write that "This reciprocal relationship continues up the hierarchy of organisation through the entire nervous system and the organism so that the term environment begins to take on its more common meaning" understood as the physical and cultural environment which we, as individuals, have the agency to act upon and change in the process of actualising the possible futures intended-towards.[45] It is thought that the main role of the prefrontal cortex, located in the frontal-most lobe of the cortex behind the eyes, is to mediate an integration of its own functions and those of other areas of the brain intended upon an object, whether that object be internal or external to the subject.[46] Laughlin, McManus, and d'Aquili claim (after Pohl, 1973; Mishkin et al., 1977) that "The prefrontal lobe seems to be a multimodal association area, most likely concerned with 'egocentric spatial orientation' *toward*, and integration of information *about*, discreet events in sensorial space".[47] Daniel Dennett, supporting a similar view, is quoted in Flanagan as stating that the "key to control is the ability to track or even anticipate the important features of the environment, so all brains are, in essence, anticipation-machines".[48]

What has been suggested here is that a process similar, and likely phylogenetically related to that which drives the goal-seeking ambition of the neuron, operates at the level of the brain and nervous system (i.e. the body as a whole), and this process is responsible for driving human intentionality, the essence of our time-consciousness. In effect, "The sensorial processes constitute the world, the prefrontal processes (of the neocortex) constitute the subject rising to anticipate, meet, and cope with the world-object in intentional focus within its sensorial context".[49] Therefore, the essence of consciousness, or intentionality, can be apprehended as the qualitative effect of an innate repetitive biological mechanism arising from the level of the neuron. *Process* over *faculty* is advocated by Flanagan who states that "consciousness is taken to name a *set of processes*, not a thing or a mental faculty",[50] and thus evoking the earlier ideas of Heidegger who speaks of the process as structure:

> At issue is not the particular individuation of a concrete intentional relation but the intentional structure as such, not the concretion of lived experiences but their essential structure, not the real being of lived experience but the ideal essential being of consciousness itself, the apriori of lived experiences in the sense of the generic universal which in each case defines a class of lived experience or its structural contexture.[51]

So far I have attempted to provide an explanation, albeit rudimentary, for the set of processes which give rise to human intentionality, and which form the essence of consciousness, drawing on both Heidegger's model of time-consciousness and relevant neuro-biological theory. It is important to note that, though 'driven' or 'powered' biologically, the actual focuses of intentionality, those objects-intended-toward, and the quality of the subject-object relations, are not defined exclusively by biology, but rather involve levels of individual, social, and cultural agency in their formation. In the next part of this section, I will start by addressing the apparent motivations of the client-patrons for commissioning minarets, and proceed from there to explore the acquisition of a 'Master's level of intentionality' by drawing comparisons with the acquisition of *ihsan*, the third dimension of Islamic spirituality.

Part 2: A Patron's Motivation and A Master's Intentionality

It would seem sensible in a discussion of a Master Builder's professional intentionality that I begin by briefly addressing the role of the patron since their motivation to finance a project is a necessary prerequisite for the builder's engagement in his craft. I must qualify my study by acknowledging that, in collecting information on the patrons of the Bayt al-Maswari, namely the Bayt Fahim, I did not interview members of the family directly but instead have relied on accounts offered by the builders and from other sources in Sanaʿa who have had dealings with the family. Though they did not explicitly say so, the elder al-Maswari brothers seemed adverse to the possibility of my meeting with the Bayt Fahim, and made no co-operative effort to introduce me when I expressed interest. They may have feared that my not being Muslim, or perhaps simply my being a foreign researcher, might complicate relations with their clients, or, more likely, they may have harboured suspicions about the type of information that I would seek to elicit. Whatever the actual reason was, it was never made clear to me and I chose not to pursue contacts with the Bayt Fahim. Rather, I learned about the Bayt Fahim from secondary sources, and I also conducted short interviews with the patrons of several other small projects in Sanaʿa regarding their motives for selecting particular builders and have considered this data in a more general analysis of patrons.

Early in my field studies, when I inquired into the identity of the person who had commissioned the building of the minaret for the ʿAddil Mosque, I was told by the Head Builder that it was the "Malik al-Sukkar", "The King of Sugar", a private patron who had made his fortune in Yemen's sugar industry. Standing together on top of the minaret walls, Muhammad, motioning to the building below us, told me that this same patron had also paid for the rebuilding of the mosque and for the *qaddad* plaster that was being applied to its roof. At the time of my study, there were two wealthy industrialists popularly referred to as the "Malik al-Sukkar" in Sanaʿa: Shahir ʿAbd al-Haq and Haydar Fahim. It was the latter who had paid for the building of this minaret at a cost of nearly eight million riyals (approximately sixty-four thousand American dollars), as well as for several other minarets built by the Bayt al-Maswari. I soon discovered that the Bayt Fahim had been a prominent family since before the revolution and more recently had been headed by two brothers, ʿAbdullah and Haydar. Although both were now deceased and the control of business operations remained in family hands, Muhammad continued to refer to the patron of his project as an almost 'living, breathing' Haydar Fahim. This might be best understood in relation to the fact that, according to Murata & Chittick, Muslims believe that death at the end of life is not an absolute death, but rather the dead person continues to have experiences: "hence, this is death in relation to this world, not in relation to the whole of reality".[52]

Ahmad al-Maswari ascribed the word *sadaqah* to this form of patronage being made in the name of Haydar Fahim. This term appropriately translates to 'alms' or a 'charitable gift'. Unlike *zakat* which is paid in accordance with stipulated percentages of one's income like a tax, *sadaqah* is a completely voluntary act. In discussing the subject in *The Call of the Minaret*, Kenneth Cragg states that "on the basis of Qur'anic injunction and traditional exhortation the practice of almsgiving became a basic social institution in Islam". The author has no doubt that the act of giving "seals believing" by effectively dramatising a new allegiance between the giver and their Faith. He interestingly suggests that this form of almsgiving may also "express an otherwise inarticulate repentance".[53]

This latter insight corresponds with Ahmad's account, which explained that through the act of endowing the 'Addil mosque and its religious community with a minaret, the donor was effectively atoning for his transgressions and sins, and thus ameliorating his own chances of transcending the intermediate world of the grave (*qabr*) and 'return' after his resurrection to God heaven (*sama'*):[54] the pious beneficiaries of Haydar Fahim's endowment will pray for his salvation, and God generously offers recompense in the afterlife for those who leave behind good works. In the Qur'an it is written: "God has promised those who believe and do the right forgiveness and a great reward".[55] It is interesting to note that other benefactors who commissioned minarets from the Bayt al-Maswari inscribed their acts in the stone walls of the tower's base for posterity. For example, one reads "The building of the mosque and the minaret was at the expense of al-Raji, Mercy of God be upon him". This popular reasoning is guided by passages in the *Hadith*, such as "Whoever built a mosque (for Allah's pleasure), Allah would build for him a similar place in Paradise",[56] and reflected in '*at-Taubah*' (Repentance), *sura* 9, verse 103, of the Holy Qur'an:

> Accept the offerings they make from their wealth
> in order to cleanse and purify them for progress,
> and invoke blessings upon them.
> Your blessings will surely bring them peace,
> for God hears all and knows everything.[57]

An acquaintance who worked at the time of my study for the General Organisation for the Preservation of Historic Cities in Yemen (GOPHCY) shared his own related perspectives on the Bayt Fahim with me. He told me that it is popularly believed in Sana'a that when the last Imam, Imam Muhammad al-Badr, fled the country following his overthrow in 1962, the Fahim family confiscated many of the properties that had been in his name. Allegedly, these included some of the finest buildings in the old city, and among them several *samsarahs* (warehouse/hostels for market traders). In the recent past, GOPHCY officials had been forced into reportedly difficult negotiations with Haydar Fahim in their attempt to reclaim some of these edifices for public use. In the words of my contact, the Bayt Fahim have set up charitable trusts for the erection of mosques and minarets in order to "make their money *halal* (or legitimate) in the eyes of the community". He continued: "By making a show of it, by erecting buildings for religious institutions, they're also in a position to maintain an upstanding name for themselves as 'pious' (*wara*') citizens".

The Fahim family's active interest in securing entitlements to fine institutional buildings and their ongoing patronage for the erection of new ones has a strong precedent in the dynastic histories of the nobility throughout the Middle East, as well as in the rest of the world. By way of example, Blair and Bloom writing on the *Art and Architecture of Islam* note that during the Mamluk period in Cairo, it was common practice for prominent members of wealthy families to set up charitable funds for the erection of "large and impressive ensembles [of religious buildings] in order to glorify [their] memory through architecture". Benefactors insisted that such distinguishing architectural features as the "tombs, minarets, and portals" be located along the busier public thoroughfares in order to maximise their public audience. Like the geographical situation of the 'Addil Mosque in the old and densely built-up quarter of Bir al-'Azab, the available plots of land in Islamic Cairo were irregular in shape and called for imaginative site planning which often resulted in the orientation of some elements, like the minarets, being "at some variance to Mecca".[58] As I have discussed previously, this is also the case for both minarets I worked on, however their respective orientations with the cardinal axis were the result of strategic decisions.

There are also a number of practical economic reasons for the institution of *sadaqah* and the founding of charitable trusts. Ahmad 'Ali in a footnote to his contemporary translation of the Qur'an writes that "The purpose of voluntary contributions is not only the betterment of fellow-beings but also to restrict the accumulation of wealth in a few hands, and to encourage its free flow and circulation".[59] Commissions for new buildings such as schools, hospitals, mosques and minarets, serve to channel previously concentrated wealth into the public sphere for commendable public benefit. Also of economic note, the establishment of public endowments in the form of a charitable trust, known as *waqf* (plur. *awqaf*), serves to insure the survival of a family's wealth and name since *waqf* is protected by religious law from confiscation.[60] With regards to 'family name', Meneley notes in her study in Zabid that the descendants of important families share the inherent glory of their ancestors, and "the *sharaf* (honour) of a clan is a totality of significance derived from acts accomplished by its ascendants".[61] Like the Mamluk benefactors considered by Blair and Bloom, wealthy Sana'ani families donated *waqfs* not just for pious reasons but also to safeguard these properties from being seized by the ruling authorities, or likewise, from being divided, sold, or squandered by heirs.[62] They also served to perpetuate the glory and honour of the family heritage.

At the time of their writing in the early 1980s, Serjeant and al-'Amri claim that "some two thirds of Sana'a is *waqf* of one category or another, and baths and *samsarahs* are invariably so, as are many shops."[63] Citing Dr. al-Tayyib Zayn al-'Abidin, the authors classify five types of *waqf* generally found in Sana'a, one of which closely resembles the charitable trust established by the Bayt Fahim for the erection of religious buildings. This is the *Waqf al-Wasiyy*, or 'The Trustee's Waqf', which they define as follows:

> No part of it is given to the *Awqaf* [referring to the Government Ministry set up for the administration of waqf endowments] but the donor specifies a particular pious activity, usually a mosque to be maintained. The rest of the income is distributed among the relatives according to the law of inheritance. The Ministry hardly interferes in this *waqf*, it is controlled by a trustee, *wasiyy*, usually the eldest member of the family named by the donor.[64]

Likewise, Haydar Fahim's charitable trust is administered by his heirs in his name, and not by the Ministry of Awqaf. Effectively, the minarets and other religious institutions he has endowed through the decision making of his genealogical successors become a vehicle for both benefactor and beneficiary to direct their respective intentionalities as Muslims and as citizens with a perceived status. This practice extends likewise to other notable families in Sana'a with the economic means to do so, as indicated by an inscription placed in the stone walls of a minaret base built by the al-Maswaris: "The building of the minaret is an endowment from the late 'Abd Rabihi Muhammad in the eleventh month of 1409 AH".

For Haydar, the endowment of a religious edifice constitutes 'a good work' in the eyes of the Islamic faith, and is a means by which his sins may be atoned, even after he has died. As already discussed, architectural patronage also serves to establish and perpetuate the family name and its associated status. Interestingly, the al-Ihsan minaret, also paid for by the Bayt Fahim who are recognised as supporters of the current People's General Congress political leadership, was located in the Madinat al-Asbahi, a modern city neighbourhood of reputedly staunch supporters of the conservative Islah Reform Party. This decision may be interpreted as both a means to invoke prayer for Haydar from what is perceived to be a more pious religious community, as well as a political tactic to secure more amicable relations between the enterprising Fahim Family and the potential usurpers of (or at least, the most influential lobby group which challenges) the present government. These religious and secular ambitions, the two closely entwined, clearly demonstrate the overarching temporal framework of intentionality. The past, be it perceived as glorious or riddled with mendacious acts (or more likely both), informs a present decision to establish a *waqf* ultimately aimed at manipulating a future intended-towards, even a future that Haydar will not be 'present' here on earth for.

Muhammad al-Maswari's direct reference to the deceased Haydar as the patron of the minarets he builds may be reconsidered at this point as serving a dual purpose. Firstly, Muhammad is acknowledging Haydar as the original source of the trust in a factual sense, but secondly, and perhaps more importantly, his endorsement is also a form of praise for the good works left by Haydar. Therefore Muhammad's actions may be understood as his performance as a beneficiary of *sadaqah* in accordance with the *sura* 9:103 so that he may "cleanse and purify" his deceased patron "for progress" from the grave to heaven. While seemingly altruistic, his role may be interpreted simultaneously as an enactment of his intention to fulfil his religious duties and perhaps atone for his own transgressions, as well as an instrument for constituting and reconstituting his own identity and status as a respectable Master of his trade and moral agent of Sanaʿani Islamic society. Titus Burckhardt, remarking upon a saying of the Prophet – 'The value of actions is only through their intentions' – notes that "the inward attitude [intention] is wedded to the formal quality of the rite [action] ... and the act transcends the domain of the individual soul".[65]

Although I could not certify his claims in any 'factual' sense based on my observations at the site, Muhammad is the only member of the building team along with the aspiring *thanaʾa*, Gayyid, who professed to pray five times a day as required by Islam. He made concerted efforts to foster relations with male members of the religious communities where he was commissioned to work, and notably with the *faqihs* and the older, prominent residents of the neighbourhoods. Many of these men made regular visits to the sites, making an arduous climb to the top to repose and chat with the *asatiyyah*. For the wedding of his neighbour's sons, Muhammad hosted a gathering of more than one hundred and fifty men in his house, and played a highly visible role in organising the succession of events. During the *nashad* in the evening, when the musicians played and the *munshid* chanted Qurʿanic passages for the groom and guests gathered in the street outside the house, Muhammad adopted the role of disciplinarian, rattling his stick menacingly at the rambunctious children who scurried in all directions. His regulated conduct in his spiritual, secular, and professional life, whether projected though action or propositional claims, was very much invested in the production of his identity as a responsible, accomplished, and matured Master.

Serjeant and Lewcock specify two types of contracts between patrons and builders: the first is *al-mushaqah*, in which engagements and wages are determined on a daily basis; and the second, stipulating a fixed sum to be paid for a specified work, is referred to as *ʿamal muqataʾah*.[66] This latter class of construction contracts corresponds to all those procured by the Bayt al-Maswari. From the agreed sum of payment, Muhammad was required to budget for all necessary labour and material costs of the project, and to calculate a financial profit to be shared between him and his brother Ahmad. In the case of a minaret commission, the budget is determined in accordance with the height of the tower and the materials to be used, which may consist of either a stone base and brick shaft, or a complete stone structure like that for the Mosque of Khalid bin Walid. These factors are negotiated between builder and client in the initial stages of a contract, and as Muhammad confirmed, his patrons at this point in his career are already familiar with the repertoire of Bayt al-Maswari projects and they can be confident that their expectations will be realised. He continued: "If a patron proposes something inappropriate, I mean either stylistically or structurally, I try to persuade them to reconsider their choices and bring them into line with my family's expert discretion. If I can't negotiate the necessary changes to make the project acceptable, we refuse the contract. A project must be worthy of our reputation (*sum'a*)".

Likewise, during the course of a project, Ahmad and Muhammad made adjustments to the programme deemed necessary for the achievement of a structurally secure and aesthetically proportionate edifice. As already discussed, the originally stipulated height of forty-four meters for the 'Addil minaret was re-negotiated midway through its construction to be raised to almost fifty-five. As Ahmad proclaimed at the time, "she was all legs and no body!", referring to the seemingly stunted proportions of the tower. The builders had decided that if the Bayt Fahim refused to supplement their endowment, they would absorb the expense for the vertical extension themselves in order to preserve the noble memory of their deceased father and their reputation as Master craftsmen. During the erection of the al-Ihsan minaret, negotiations were also underway to increase the height of that structure, this time between the builders and the national security forces. Because of the proximity of the building site to the Presidential Palace, a maximum height of thirty-seven meters had been stipulated for security reasons by the Ministry of Defence. Again the proportions were perceived to be squat, and Muhammad managed to successfully mediate an agreement which allowed them to raise the elevation to forty-five meters. "It's a question of reputation", he declared. This extreme dedication to one's craft rooted in ideals and reputation is not uncommon amongst elite master craftsmen: in 1979, Egyptian Master Builder Hassan Fathy accepted the commission for the design of the Dar al-Islam project in Abiqiui, New Mexico, without a fee because of "his deep belief in the ideals involved in the project".[67]

At the start of a new minaret venture, the Head *Usta* selects a location to be excavated within the boundaries of the mosque property. All of the minarets by the Bayt al-Maswari which I visited were prominently situated to secure the best views of the structure along a number of sight axes. All stood as separate structures from the mosque buildings, and in many cases were adjacent to, or coextensive with, the property walls nearest the street.

As Bloom points out in his formidable study of the minaret as symbol of Islam, by the middle of the eleventh century the minaret lost its "toposensitivity": as it became progressively recognised as an Islamic sign, this tower form no longer "derived its meaning from its proximity to a mosque", and was even occasionally built "quite independent of any structure".[68] By the sixteenth century, Sinan's stylised domes and minarets "provided a means of visually marking the broad extent of the Ottoman dominion", thereby taking on the political dimension of an empire.[69] In contemporary Sanaʻa, the prominent siting of these elaborately crafted towers heralded the identities of the builder and, more significantly, the patron, and their distinctive architectural form signposts the spiritual centre of the community for whom they are supposedly erected.

The issue of quality in craftsmanship has been raised repeatedly throughout this study, focusing on the disciplinary stages undergone by the young builders which aim to inculcate proper comportment and basic skills, and the subsequent stages geared to sediment a more advanced level of understanding of the trade processes in the apprentice and the younger *asatiyyah*. The apprentice arrives at a level of 'understanding' through repeated engagement in the practice of his performance, as well as through the introduction to crafting materials, or 'making'. This latter stage of knowledge acquisition and 'making' endures beyond the apprenticeship, well into the builder's vocation as a young *usta* working under the direction of a senior Master. The senior Master typically conceives of the project 'holistically', being sufficiently accomplished in all aspects of the construction and management, and the subject of his professional trade comes to occupy the full volume of his mind. Robert Grudin writing on the processes of creativity notes that "attention to one's subject should be so generous, extended and intimate that the idea virtually inhabits the mind", becoming "a kind of world in itself, crowded with the forms and potentialities that tie it to the rest of experience".[70] Charles Taylor, in his essay on *Hegel's Philosophy of Mind*, writes that, for Hegel

> the goal of spirit is clear, self conscious understanding. But the struggle to attain this is just the struggle to formulate it in an adequate medium ... The only fully adequate form is conceptual thought, which allows both transparency and full reflective awareness. But attaining our formulation in this medium is the result of a long struggle. It is an achievement; and one which builds on and requires the formulation in the other, less adequate media,

which Hegel identifies in ascending order of adequacy as art and religion.[71] It was Muhammad's disciplined immersion in both of these (as craftsman and pious Muslim) that facilitated his transcendence to a higher level of self-conscious understanding as a Master of his trade.

This expert level of absorption and concentration was evident to me while observing Muhammad al-Maswari at work. His entire performance was deeply entrenched in an intentionality toward the project as a whole, and not simply in his own particular task at hand. Like a symphonic conductor, he was continuously aware of all members of his building team, monitoring their performance and bringing them back into line with his own standards and vision when necessary. His demanding demeanour was tempered by a composure and fairness of judgement, earning him a great deal of respect from those that worked for him. All of his actions were precisely calculated, enabling him to move with great economy and distinguishing grace, maintaining a regular, even tempo in the rhythm of his execution. For example, when assembling the exterior facing of masonry on the walls of the minaret base, he meticulously set each cut stone into its perfect position by repeatedly verifying its horizontality and verticality with his simple level and plumb-line, and gently tapping it into place with the dull end of his adze. Systematically, he then proceeded to finger the oozing grout around every joint, almost caressing the face of the stone with his other hand while eyeing it intensely to verify its position in relation with the other stones in the course.

Muhammad's relationship with the building materials he employed in all his projects seemed to go beyond simply one of objective knowledge, and he engaged with them at an almost intimate level. He intrinsically knew how to work *with* his materials, aware of their potentials and limitations, not just generally, but of each individual brick and stone which was scrutinised by his gaze. Bricks of a certain quality were kept for carving and others were relegated to the infill of walls; dressed stone was accepted or rejected depending on its colouring, its solidity, and the direction of its grain. In *Islamic Art and Spirituality*, Nasr observes that traditional Muslim craftsmen "had a profound sense of the nature of materials" with which they dealt. "Stone was always treated as stone and brick as brick" and the materials were masterfully integrated "into a whole reflecting the ethos of Islamic art".[72] This 'sense of materials masterfully integrated' unquestionably characterised the works directed by Muhammad al-Maswari: his knowledge and craftsmanship transformed an 'act' into the 'art' of building.

In contrast to the zeal and dexterity applied to the perfection of his work, Muhammad, like the other Master Builders I encountered in Sanaʻa, placed little emphasis on improving the rudimentary tools used by him and his builders. Adze and hammer heads frequently disengaged from their wooden handles, tape measures were rusted, and the rubber buckets used to transport mortar and rubble between the top and bottom were riveted with holes and cracks. Interestingly, Titus Burckhardt notes in relation to this seeming neglect of the tools and instruments of the trade that the "attitude is at least partially explicable by the very acute awareness that a Muslim has of the ephemerality of things; art has something provisional about it ... and the craftsman's greatest achievement is the mastery he gains over himself".[73]

The Master Builder and Ihsan: a balance between reason and imagination

In following similar ideas set forth by both Burckhardt and Nasr,[74] I will draw comparisons between this level of mastery achieved by the accomplished craftsman, like Muhammad al-Maswari, and the highest level of spiritual realisation attained by the pious Muslim. Firstly, this highest level of spirituality is the third and final dimension of Islamic enlightenment, and is defined as *ihsan*, literally 'the performance of good deeds'. *Ihsan* follows on from *islam*, or submission, to which I drew parallels with the first stage on the career path of a builder, labouring; and *iman*, or faith (in what one is submitting oneself to), which I briefly compared to the apprentice's increased understanding of his craft through the processes of 'making'. According to Nasr, Sufi Masters have historically defined the ultimate goal of the noviciate's journey toward spiritual enlightenment by a well known *Hadith* of the Prophet which explains that "*Ihsan* is to adore Allah as though thou didst see him, and if thou doest not see him He nonetheless sees thee".[75] Maulana Muhammad Ali explains in his *Manual of Hadith* that "*ihsan* is not a technical term and indicates the state of sincerity in one's conviction [read *iman*] or practice [read *islam*]".[76] The notion of *ihsan* as a state of sincerity, or *ikhlas*, is also supported by Nasr who writes: "to possess this sincerity is to make one's religion central and to try to penetrate into its inner meaning with all one's being". When *ihsan* transforms *iman* (faith), it becomes "an illuminative knowledge",[77] like that possessed by the Master Builder.

This 'illuminative knowledge' resulting from *ihsan* is what Murata and Chittick associate with "doing what is beautiful" (from the verb *hasuna*, 'to be beautiful') with "a focus on human intentionality".[78] In the meticulous and graceful performance of his task, whether it was the setting of a stone or a negotiations between himself and the builders under his tutelage, Muhammad's intention was fully focused on creating "what is beautiful" in his craft. The creative insight involved in his ongoing decision making processes regarding materials, decorative patterns, proportions, and project management was predicated upon his own practical training, experience, and prolonged concentration as a craftsman. Grudin suggests that this creative insight, or inspiration, is the result of "the combination of an active principle – hard-earned expertise – with a passive principle – unencumbered and trustful receptivity", the latter which is ultimately attained when one's mind is fully absorbed in the subject-intended-toward.[79] This intrinsic bi-polar quality of creativity corresponds to the Islamic acknowledgement that the principle of reason which actively discerns difference, and is associated with *tanzih*, or 'God's incomparability with anything else', must be balanced with the connectedness wrought by imagination (*takhayyul*), associated with *tashbih*, or 'God's similarity to everything'.[80] Together the properties of reason and imagination form a balanced whole, just as *tanzih* and *tashbih* constitute the two poles of *tawhid*, or the 'Oneness of God'.

Muhammad's focused intentionality toward his trade had been wrought by the disciplinary measures of his training to which he was subject through his own acts of self-subjectification. It enabled him to engage in an "unencumbered and trustful receptivity" to imaginatively conceptualise the connectedness between the varying aspects of the building project. His holistic vision encompassed the physical scales of *objects-as-entities*, ranging from the brick to the tower, and the spatial relations between these; and the web of his own emotional and professional relations to his clients and builders. Likewise, his disciplined intentionality, as producer of his time-consciousness, clearly operated at varying scales of temporal construction: from the immediacy involved in the decisions of the conceived 'now' while engaged in the associated practices of the act of building; to a conceptualised chronological delineation of the project programme; and onward to 'long-term' intentions informed by concepts of the structural integrity of the edifice, the 'memory of his father', and the 'reputation of his building house'. Muhammad's expert capacity in formulating connections between the physical, spatial, emotional, and temporal aspects of his work is closely related to the 'illuminative knowledge' of the Sufi Master which enables him to spiritually conceive of *tawhid*.

In warning, however, Burckhardt records in *Letters of a Sufi Master* that "It is the nature of imagination [which enables one to conceive of connectedness] that if you do not bring it under your yoke, that is, impose your way of thinking upon it, it will inevitably dominate you and impose its own way on you".[81] The passage stresses the need to stabilise imagination with the 'yoke of reason' (*ʿaql*), or balance *tashbih* with *tanzih*. Importantly, what is suggested here is that the 'free-flowing' processes of imagination are not, in themselves, under the control of a conscious agency. On the other hand, as I have determined, the processes of *arriving at* that "unencumbered and trustful" state are propelled by the disciplined, conscious intentionality of an agent, like the processes of *reasoning*. The qualitative properties of Muhammad's *reasoning* which guided his decisions and disciplined his creative imagination were rooted in a number of considerations including what could be practically achieved with the tools and materials at hand; what could be accomplished by his team in view of their capacities, both physically and vocationally; what was negotiated as desirable by his clients in terms of scale, style, and quality of construction; and, notably, what was ideally conceived as perpetuating 'tradition' in Sanaʿani architecture. For example, Muhammad's *reasoning* in relation to the latter harnessed his creative imagination for minaret design by focusing his creative intentions within the conceptual boundaries of what has been historically ordained 'traditional' in the city.

In short, an agent's reason – the cognitive capacity to discern and categorise *objects-intended-toward* – is actively engaged with imagination in order to 'stabilise' the concepts produced there and to render them objects of intentional focus. An object of intentional focus can be acted upon and manipulated conceptually, not unlike the *object-as-entity* produced both physically and conceptually in the *processes of making*. It may be useful at this point to ponder the parallels between the abstractness of 'spatial relations' and of 'imagination'; and between the concreteness produced by the 'processes of making' and by the 'processes of reason'. The *processes of making*, as determined in the previous chapter, give rise to the re-conceptualisation of *spatial-relations-between-objects* as newly formed *objects-as-entities*; in a similar manner, the *processes of reason* serve to 'stabilise' and 'define' points in the web of conceptual connections produced by *imagination*. In the case of the Master craftsman, such as Muhammad, the higher faculty of cognition responsible for producing conceptual metarepresentations is characterised by the balance between the competencies of his imagination and reason, and this mastered 'balance' is the *sine qua non* of his expert status.

Finally, it should be noted that not all aspects of the Master craftsman's professional knowledge can be transmitted to those under his tutelage through instruction. Students may be taught to perform craft-related skills with a sufficient degree of discipline, as the Muslim child acquires a mastery over the performative aspect of their religious duties (*islam*); and those motivated to remain professionally in the trade may adopt a certain level of understanding about their vocation through their training (for example in the *process of making*), just as motivated Muslims engaged in their respective spiritual vocation will progressively acquire an understanding and faith (*iman*) in Islam. However, the *usta*'s mastery over his expert knowledge, distinguished by the *balance* between his cognitive capacities for imagination (*takhayyul*) and reason (*'aql*), is not amenable to the same type of teaching-learning transference, and is highly dependent on the inherent qualities and character of the apprentice. Like the Sufi noviciate striving to attain *ihsan* in his spiritual quest of *being*, the young builder must first demonstrate an 'inward vocation' predicated upon his *aptitudes*, *motivations* and *aspirations* toward attaining this highly focused level of intentionality on "doing what is beautiful" in his craft.[82]

Concluding Remarks: Architects and Traditional Builders

At the time of my field study, there was very limited work for professional architects in Yemen, especially for those outside the major urban centres of

Sanaʿa, Hodeidah, Taizz, and Aden. This was not surprising considering the long and formidable history of *architecture without architects* in this country. The architects whom I knew from the university, many with graduate degrees from outside Yemen, complained that if they did receive a commission to design and build, their role and power in the project was frustratingly restricted, and this was further aggravated by the meagre salaries they could command in the present economy. The world money market value of a typical fee for the design of an urban villa in Sanaʿa had dropped from approximately four thousand American dollars in the 1980s to between one hundred-and-sixty and four hundred dollars in 1997. Many clients were now only willing to pay for a simple set of drawings that consisted of floor plans without any elevations or working drawings. Typically, these drawings were handed over by the client to a traditional Master Builder, or *usta*, who would be fully entrusted with the execution of the building.

With a set of drawings that amounted to little more than an architect's exercise in floor-plate planning, the traditional builders was normally given free reign on the choice of materials and on the overall architectural expression and aesthetics of the building. Architects frequently accused the builders of making 'unauthorised' changes to their designs, or of neglecting to follow the plans altogether, but their positions were further thwarted by a number of barriers. When disputes did arise, clients often subscribed to the authority of the trusted traditional builder for the ultimate decisions. In practical terms, there was no means of legal recourse for the architect, and in the bureaucratic confusion of the justice system, the designs he/she produced found no protection under copyright laws. In a country where formal architectural training is a recent phenomenon, the general public has largely retained their trust in the tactile expertise of the *usta* over and above the abstract 'paper ideas' of the architect. In an interview recorded by Hatim al-Sabahi, the elderly Master Builder al-Haj ʿAbdullah al-Rawdhi recalled from his own experiences that "The owner of the house decides in agreement with the Master Builder on the arrangement of floors and rooms, according to the dimensions of the site and without having a plan".[83]

Interestingly, and ironically, despite constant complaints about the intervention of traditional builders in their projects and the power accorded to them, many of the architects I knew publicly lamented the prophesied demise of the *usta*'s traditional trade, and feared that the country would lose its distinctive architectural heritage to technological progress. Personally, I saw no proof of this apparent 'demise' as there was ample building activity throughout both the country as a whole and the city of Sanaʿa by so-called 'traditional builders' (*banna'un taqlidiyyun*), and many of these projects were being commandeered by accomplished *asatiyyah* assisted by teams of village labourers and apprentices. As I indicated in the last chapter, I strongly suspected that the 'lamenting' by these architects was not actually in response to any genuine situation, but rather it masked their unprofessed 'hopes', not 'fears', of what the situation *would eventually become*. Hypothetically, a demise of the traditional building trade would signal the ascent of the architect's professional status, and their own formalised technical knowledge (*'ilm*) would pervade the social arena of expert discourse on building.

In the mean time, the traditional Sana'ani Master Builder has managed to maintain the upper hand in the struggle between *usta* and architect, though many concessions have been made since the 1970s marking an ongoing re-negotiation of the power share. As I have demonstrated in the previous three chapters, the content of the Master Builder's knowledge is elusive by nature of its inherent teaching-learning processes which are embodied in the apprenticeship. Knowledge is sedimented in the performance, the understanding, and ultimately the intentions of the builder through discipline, mimicry and practice, and a hard-earned focus of concentration and reflectivity. This is not an objectified knowledge (*'ilm*) that can be acquired by following a formulaic recipe of procedures, but rather can be properly characterised as *ma'arifa*, a deeply personal knowledge acquired by following the path (*tariqah*). Previously, it had not been written down and its particular educational processes had not been objectified as a social fact which could be reflected upon, discussed, and disputed in either quantifiable or qualitative terms. The common response issued by the *usta* when asked about his knowledge is that "its all in my head" ("*kull fi ra'si*"), and this is generally accepted by his public without further questioning, and reproduced in their own sustained beliefs about the 'inherent qualities and powers' of master craftsmen. As I have shown, precisely because the expert knowledge is not objectified, the traditional builder is not capable of explaining *how he knows*, and arguably, to some extent, he is unwilling to explain and is fully aware of the power wielded by the ambiguity in his responses.

The *usta* has remained the sole custodian of the reproduction of his trade largely because the nature of the teaching-learning processes and the master-apprentice relations have remained elusive to objectified scrutiny and reproduction by outside agents. In conclusion, I return once again to my discussions with architect Hatim al-Sabahi in which he proposed that the crafts related to the traditional building industry be institutionalised in the form of trade schools in order to safeguard them from being obliterated by the wave of modernism and radical reform in the country.[84] From my own studies with traditional builders in both Nigeria and Yemen, I have concluded that the specific teaching-learning processes are not amenable to such a formalised method of instruction. Though the crafts may be reproduced, the trade, as a distinctive set of human relations and method for transferring not only technical, but personal knowledge, cannot. It is the master-apprentice relation that is the distinguishing hallmark of the trade, and not the actual objects which are produced by it. In his study on *Traditions as Truth and Communication*, Boyer concludes that "casting traditionally created truths in the written format is the first step on a slippery slope that may lead to the discussion and contestation of these truths".[85] With a developing democracy in Yemen, and a slowly increasing demand by the public for a democratic access to knowledge, the objectification of these 'truths' in written propositional forms seems imminent. After all, my ability as a Western researcher and *nasrani* to reside in once-forbidden Yemen and write this down is part of that democratic process.

CONCLUSION

This study has focussed on a particular team of traditional Yemeni builders, the Bayt al-Maswari, who have specialised during the past two decades in the construction of mosque minarets in and around the capital city of Sana'a. In consideration of the professional statuses, roles and individual personalities of several of the builders, and through analysis of the social and political relations that existed amongst them during the time that I worked as a labourer, I have aimed to elucidate the manner in which the young builders gradually mastered both the self discipline and trade skills necessary for their eventual recognition by the Masters as competent craftsmen. In doing so, I have sought to better understand several salient cognitive operations which were exercised and developed by the particular teaching-learning processes involved in the traditional builder's apprenticeship. More specifically, those highlighted in this study have been the labourer's obedience and regulated mind-body comportment achieved through his subjection, and self-subjectification, to the disciplinary apparatuses on the work site and in his living environment; the apprentice's capacity to observe, practice and reproduce trade skills which involve the mastery of spatial judgements, and which are cultivated through the act of 'making'; and finally, the honing of the Master's intentionality which imbues and directs the conceptual coordination of the numerous aspects involved in the trade (including design, economics and resource management, client relations, and the capacity to conceptualise the project at various physical scales).

In the first instance regarding the training of the labourer, the emphasis on inculcating obedience and disciplined comportment, devoid of any explicit understanding of the tasks being exercised or their relation to the overall project, was compared with *islam*, the first stage in the socialisation of a Muslim child. Learning to perform prayers and ablutions, and to recite the Qur'an, marked the early stages of an Islamic education and the development of a moral self. As vom Bruck notes, the nature of this religious learning was "not aimed at *understanding* in a sense of 'Western' pedagogics" but rather was meant to bring "children's minds and impulses under control ... and thereby make the pupils people of Islam".[1]

A traditional Islamic education, like the training of a labourer at the building site, was characterised by "rigorous discipline" and a "lack of explicit explanation of memorized material".[2] In his Moroccan study, Eickelman observed that these features were "congruent with the concept of essentially fixed [Islamic] knowledge ... and the associated concept of 'reason'".[3] Likewise, the tasks and duties performed repetitiously by the young building labourers comprised a fixed and limited corpus which was not subject to interpretation, but rather required a domestication of reason, 'aql, and a submission to an economy of bodily performance. Discipline was an integral part of the socialisation of the builder as both a labourer and a 'moral' person who was integrated as an efficient member of the team. The history of the individual builder ultimately became nothing other than "a certain specification of the collective history of his group", and his individual dispositions become regulated as a mere structural variant of the group *habitus*.[4] What was important at this early stage in the training was learning how to 'act', not yet how to 'understand'.

It was noted that the labourer, in order to successfully transcend his lowly status and be delegated increasing responsibility and a greater role in the construction process, necessarily had to possess a healthy measure of skilled aptitude, self motivation, and an aspiration to remain within the trade and to become a Master. These qualities divided those who would be considered for further training by the Masters from those who were either incompetent or those complacent to work for a base wage without incurring the burden of further responsibility or commitment. Not all labourers had the endurance, the self discipline, or the will to become craftsmen. Many maintained professional interests outside the trade and laboured solely to accumulate capital to support themselves and their families back home in the village. Even the youngest generation of al-Maswaris, presently the fourth generation of professional builders, demonstrated limited interest in pursuing careers in the trade. In comparison with the former generations of their family and with the predominantly rural stock of labourers, these young urban dwellers are relatively affluent and have access to formal education and a variety of employment options. The family heads, while hoping that their bloodline will continue to command and control the expert trade knowledge, also support the tangential ambitions of their progeny. Like the traditional fishermen of Shihr on the Arabian Sea coast studied by Sylvaine Camelin, the Masters consider their work "épuisant, dangereux et qui les maintient tout de même toujours au bas de l'échelle sociale, même si leur revenu est actuellement tout â fait correcte".[5] Though the Master Builders are not at the bottom of the social scale, and in fact command a great deal of respect for their expertise, they recognise their children's opportunities to transcend their status (and the physical burden of their manual labour) by becoming doctors, lawyers, or civil servants.

In fact, only a small minority of the labourers showed any promise of becoming future Master Builders who would be dedicated to perpetuating the high level of trade quality demanded by the Bayt al-Maswari. At the time of my study, all but one of these were from outside the family. They had all acquired varying degrees of mastery over their comportment and had successfully incorporated the performance of their tasks into an unreflected form of bodily knowledge. To do so, they had implemented a number of physical factors into an effective system of mnemonics for the economic reproduction of their performance. These included their own positionality, both within the hierarchy of the work team and their physical placement on the site in relation to others; and the tempo of the work, often regulated by the rhythm of *hajl*s (work songs), and the counting of stairs and measurement of space in their allotted work stations. As Weir notes, "Yemenis are acutely sensitive to the spatial dispositions and relations of individual and groups".[6] Because so few demonstrated the motivation to discipline their own performance to this notable degree, and to likewise engage in monitoring the behaviour and actions of those working around them, mnemonic possession could be considered, as Eickelman rightly suggests, a form of cultural capital.[7]

The most coveted position for the ambitious labourers was to work next to the Master Builders and to engage in the processes of 'making', which normally included an introduction to, and a frequent engagement in, the carving of (and occasionally the setting of) *ajurr* bricks. At this stage a degree of understanding regarding the trade and one's role within it developed, and I compared this with *iman*, the second level in the path towards Islamic spiritual enlightenment. However, once construction of the brick shaft above the stone base commenced, there was space and necessity for only one apprentice at any given time. This incited an acute competition between candidates, and periodic changes initiated by the Masters based on a politics of favour spawned a perpetual state of insecurity amongst them. When recognised and appointed, work at the top of the minaret staircase included regulating the arrival of supplies, preparing tools and materials, and catering to the needs of the sometimes two, three, or four *asatiyyah* who set the courses of bricks while squatting on top of the walls. The labourer in this pivotal position was subject to a considerable degree of mostly verbal, but occasionally physical, abuse. The uncertainty of their status, in combination with the abuse inflicted by the senior builders and the stress of an expanded repertoire of responsibilities, resulted in anxiety which regularly became manifest in hostilities directed toward subordinate colleagues and particularly toward potential competitors.

The fluidity of their position amongst the hierarchy of labourers was also interestingly reflected in the absence of a distinct term used by the Sana'ani builders to designate the status of 'apprentice' or to qualify the role of 'apprenticeship'. Rather, they were both conceived of as 'process' and not as part of a rigid morphology of rank. The apprentice was simply referred to as *shaqi*, 'labourer', and less commonly as *sabi al-usta*, 'lad of the Master'. Likewise, Camelin notes that the apprentices in Shihr's fishing trade are likewise accorded titles which refer to their assigned functions and not to their status as 'learners'. For instance *farram* designates 'the apprentice who must collect hand-size pebbles from the beach, locally referred to as *farm*'; *mowal qiddam* designates 'the apprentice who sits in a place (*mowal* in this context) in the front (*qiddam*) of the boat'; and *mowal suka al-digal* designates 'the apprentice who occupies the place (*mowal*) on a bench (*suka*) where the mast (*al-digal*) is located'.[8] It is important to note that in the case of both the builders and the Shihr fishermen, although there were no words commonly used to describe the status of the trainee as a 'trainee' or the particular training process at this level, the status and role were nevertheless recognised, aspired to, and reproduced. As I have argued, to "recognise, aspire to, and to be reproduce by agents" requires an intentionality which equates to a conscious awareness of the concept or thought, and it was demonstrated that there are distinct kinds of thought which are largely independent of language. It was thus deemed necessary to go beyond language in my investigation of conceptualisation, and to adopt a modular approach to *thinking about thinking*. In the given case, 'process' is similar to a propositional statement in that it may be thought about, scrutinised and manipulated, but importantly it may be distinguished from propositional thought in that it functions without the use of language.

A modular approach to understanding the mind was elaborated in chapter three, and Jackendoff and Landau's thesis on the relation between language and spatial cognition was introduced. They concluded in their study that there exists an evident correlation between the split in the expressive power of language, namely between the ability to describe the *what* of an object versus the *where* of an object, and the anatomical bifurcation for spatial representation in thought.[9] My study has sought to expand their window onto human spatial cognition from one which was primarily language-based to include the processes of making. My focus on action versus verbal communication was necessitated by the fact that there was often little or no spoken directives issued in the training of the traditional builders, with the exception of reprimands and abuses. This seemingly widespread characteristic of apprenticeship is noted repeatedly by those who study this distinctive type of teaching-learning arrangement. It was therefore essential to develop an understanding of how conceptual representations and bodily knowledge interface dialectically, one informing and modifying the content of the other.

The builder's apprenticeship served to enhance concepts and judgements regarding space and assembly through training, practice, and inhabiting the 'processes of making' which largely defined his role and position. The labourer was typically initiated into the process of making with the carving of his first brick, and from there the scale of *making* was gradually increased. As the traditional builder mastered increasingly complex spatial relations which were inherent in the increasing scale of assemblies, these abstract spatial relations were conceptually translated and simplified into *known objects* (i.e. the brick; the wall; the decorative brick patterns; the staircase; the base, the shaft, the dome; and, ultimately, the minaret-as-entity). Consequently, these objects could be manipulated, both conceptually and physically, in an incremental manner in order to move between scales of different magnitude, from the micro to the macro, over the course of the apprenticeship. As I explained, the propensity to conceptually reconstrue spatial relations into known objects demonstrated an entropic process of cognition which guided the builder's understanding of an object from one of abstract spatial relations toward a categorised description of the object-as-entity. These categorical concepts served as discreet building blocks of knowledge over which the mind of the builder exerted power as 'maker', manipulating and re-configuring relations between these objects of knowledge to create new objects of knowledge. This process of manipulating conceptual building blocks was in dialectical relation with the physical processes of making, since, ultimately, the *knowledge* and the *making* were demonstrated to be one and the same.

Access to the practices of making and to the trade knowledge were shown to be tightly controlled, both consciously and unconsciously, by the chief builders. Confronted, however, by a crisis of insufficient representation of younger family members in the trade, Mohammad and Ahmed al-Maswari elected to reproduce their trade skills by training unrelated agents. While preserving craft knowledge and modes of production, they were simultaneously engaged in the erosion of their family's own controlling position within the expert discourse. With regards to the challenge of locating the source and the content of the *usta*'s trade knowledge, it was concluded firstly that, because of the non-propositional nature of the builder's expertise, it was difficult, if not impossible, for the traditional builder to explain 'how' he knows. To do so necessarily involved a translation of his *knowing how* into another domain, namely a propositional one encoded by language, and therefore suffered a substantial loss of specificity and content.[10] Secondly, the notion of a deliberate secrecy exercised by the *usta* was postulated, but this obstruction was contextualised within a framework of intentionality. The reluctance to qualify sources and methods of knowledge, and the frequently misleading or embellished responses, were reconsidered as integral to a 'customised discourse' aimed at producing and reproducing 'customised persons'.

In order to discuss the 'intentionality' of the traditional builder as a customised person, I developed a working model in chapter four which was rooted in a combined understanding of consciousness and experienced temporality, based upon a Heideggerian theory of time-consciousness and drawing upon relevant neuro-biological hypotheses. I argued that all decision making was driven by intentionality, and that, as a process, intentionality effectively structures our human sense of temporality and the *present-ness* which anchors our experienced identities. Though likely 'driven' by biological factors, it was emphasised that the actual focuses of human intentionality, those objects-intended-toward, and the qualitative aspects of subject-object relations, are invested with a significant measure of the individual's socially and culturally impacted agency. The agency of the Master Builder, which directs and focuses the processes of his intentionality, was brought under control through the disciplining of his reason during his own apprenticeship, and was honed by his practice, dedication, and experience in the trade. In this respect, the master craftsman's intentionality was compared to *ihsan*, the third and final stage in Islamic spirituality. At this level, one's entire 'intentionality' has become directed toward, and absorbed in, 'doing what is beautiful' in one's vocation. A controlled and focused intentionality imbues both one's 'submission', *islam*, to the requisite practices and performances, and one's understanding of the knowledge, premised on an unquestioning 'faith', *iman*.

The chief *usta*'s creative insight as a builder was recognised to be the result of a combination of an active principle, namely his hard-earned expert reason, and the unencumbered receptivity of his imaginative processes. Together the properties of reason and imagination form a balanced whole, and a mastered 'balance' is the *sine qua non* of the builder's expert status. His focussed intentionality permitted him to fully engage in an unencumbered and trustful receptivity to imaginatively conceptualise the connectedness between the physical, spatial, emotional, and temporal aspects of the building projects. His holistic vision encompassed the physical scales of objects-as-entities, ranging from the brick to the tower; the spatial relations between these; and the web of his own emotional and professional relations to his clients, builders, and family members. Likewise, his disciplined intentionality, as producer of his time-consciousness, clearly operated at varying scales of temporal construction, from the immediacy of site decisions, to the planning of projects, to the 'long-term' intentions informed by the constitution and reconstitution of the reputation of his *bayt* (building house), and his reputation as a 'moral' person.

The *usta*'s mastery over his expert knowledge, distinguished by the balance between his cognitive capacities for imagination and reason, is not amenable to the type of transference as are craft-related skills. Rather, the acquisition of this attribute is highly dependent on the inherent qualities of the apprentice, as well as on the intensity of the relation which evolves between the teacher and pupil during the course of the apprenticeship. The young builder must possess an inward vocation which is predicated upon his skilled aptitudes, motivations to secure a position, and aspirations toward attaining the highly focussed level of intentionality for 'doing what is beautiful' in his craft.

As I concluded in the final chapter, the *usta* has remained the sole custodian of the production and reproduction of his trade knowledge, and more particularly of the distinctive teaching-learning processes that are cultivated between Master and apprentice. It is not the particular objects produced by the trade (i.e *ajurr* brick minarets with traditional decorative patterning), but rather the Master-apprentice relation which distinguishes traditional craft production with a special form of knowledge. The apprenticeship moulds not only the young minaret maker's trade related skills, but importantly the moral being of his person. I have suggested on several occasions in the course of this study that the most serious threat to the perpetuation of this form of education is not the introduction of changes to materials, technologies, or consumer tastes, nor is it the presence of a university trained community of indigenous architects and engineers, but it is the advent of an emerging, fledgling democracy in Yemen. The country maintains a constitution which articulates some "basic rights and freedoms, [an] elected legislature, a certain degree of civic and political pluralism, [and] courts that tend to protect the law", which together offer Yemen the greatest potential for democratisation among Arab countries in the medium term.[11]

In less than three decades since the end of the civil war in North Yemen and the installation of a republican government, the country has witnessed the opening of its borders to the flow of peoples and ideas; the construction of an extensive road network which connects cities, villages and ports; participation in the global economy; two nation-wide democratic elections; and national and international communication links via the radio, television, fax, and more recently the internet. At the time of my study, an old man sat cross-legged on the side walk of busy Abdul Mogni Street, crowded with shoppers, with three plastic view masters neatly displayed on the ground in front of him, a leather satchel of view master slides by his side, and a cardboard sign inviting passers-by to "See the World. Just 10 riyals!"; in 1998, the first internet café opened in Sanaʿa; and middle class Yemenis continued to enrol in English, French and Japanese language classes. These are only a few examples which indicate the current changes, re-definitions, and expansions of political and public spaces, and the ripening capacity for a more individually-based (understood in Western terms) exploration and enquiry.

Carapico notes that neither foreign influences nor the ruling regimes of the country to date are responsible for promoting democratisation or liberalisation in Yemen, and she concludes that the "relative pluralism and tolerance of the Yemeni polity in the mid 1990s" must be explained by cumulative home-grown civic pressures.[12] It has been argued that Western governments have in fact turned their backs on these local initiatives in their bids to appease regional Arab powers hostile to democratic developments in Yemen and thereby protect their own "steady relatively cheap supply of oil" coming from these non-democratic states.[13] Despite prevalent assumptions found in much writing on Arab political culture that 'Arab civil society' is an oxymoron',[14] and convictions that Islam constitutes a rival form of social order to civil society,[15] Carapico contends that "the extent and range of activism in Yemen challenges stereotypes of inherent conservatism usually attributed to tribalism and Islam".[16] She convincingly argues that both tribalism and religion supply ideological appeal and moral economies, as well as systems of law "capable of functioning in the absence of a state, and each contains customary mechanisms for creating public social capital".[17] Their inherent ideas of "equality, election, consultation, [and] rights of dissent",[18] and their provision of mechanisms which operate in that contested civic space lying between the state and private sectors of Yemeni society, constitute the basis of civil society and a democratic process.

In attempt to liberate concepts of civil society from their anchorage in Western European Enlightenment ideals, Beckman observes that there exists a variety of 'civil societies' which do not necessarily support democratisation in a liberal sense, and he asserts that we should also be addressing the social basis of political regimes which may allow for "patriarchal, Islamic, communist, and fascist civil societies".[19] Though Yemen's civic space is nurturing the seeds for democracy, it must be recognised that the religious and tribal mechanisms operating there are likely to give rise to democratic processes notably different from European and American models, and should be assessed according to an expanded, and culturally relevant, set of criteria. In the absence of a full-blown democracy in South Arabia, other issues may be of more significance than the overtly political participation of Yemenis in "polls and referenda whose outcome is known in advance".[20]

A pertinent issue in studies of traditional craft and apprenticeship, and one which invites further investigation, is the increased access to knowledges and information now available to a sizable community in the country. Though presently this community represents a privileged class of Yemeni society, it is highly probable that the demand for access to knowledges and information will infiltrate all social classes and institutions, and will pose a challenge to existing educational systems including the distinctive teaching-learning mechanisms and Master-apprentice relations which characterise traditional apprenticeship. Perhaps most interesting to consider in further studies would not be the type or quantity of the knowledge and information being sought and exchanged, but rather the transformations in popular attitudes toward knowledge and the manner by which it is demanded and acquired. Predictably, greater access to knowledges, complimented by a greater tolerance of enquiry, will empower a future general public, including young builders, to question and challenge the once sacred realms of the Master craftsman's expert discourse. Changes in perspectives toward authority, and increased opportunity to investigate, discuss and challenge traditionally guarded realms of knowledge will eventually lead to the demystification of the Master Builder's expertise in the minds of the public. Revelations about '*how* he knows' and '*what* he knows' will serve to undermine his real power and consequently threaten the existence of the distinctive teaching-learning method and inter-personal relations supplied by the traditional apprenticeship system. This does not necessarily equate to the eradication of the building crafts as a mode of reproducing 'traditional' objects and buildings, but the very nature of the trade knowledge, as a marriage of both technical and deeply personal knowledges (*'ilm* and *ma'arifa*), as well as the manner in which it is both passed along and embodied, will be significantly and perhaps irreparably modified.

GLOSSARY OF ARABIC TERMS

'abasari	light-weight, beige stone used for constructing the central column of the minaret and the interior wall faces
'adhan	the call to prayer
ajurr	kiln baked clay brick
'amm	uncle, or title of respect conferred upon elders
ansaf	half (of an *ajurr* brick)
'aql	(faculty of) reason
asabi'	tamarisk branches used to support spiral stairs
'athl	tamarisk
bayt	house, also used with kinship reference, i.e. 'the House of ...'
bustan	garden
daka	stair tread
daqiqa	stair
dawwar	balcony
dhabr habash	a corner stone with two dressed faces
dhurah	sorghum
dibla	diamond-shaped relief in the decorative brickwork
diwan	reception room
djinn (jinn)	invisible beings
fa'as	axe-like tool for chiseling stone
faqih	expert in Islamic jurisprudence
furud	religious studies
gandal	courses of wood between courses of masonry
gharbala	to sieve or sift sand, gravel
ghulam al-mumahan	vocational servant, or apprentice
haba (s.) *habat* (pl.)	brick coursing
habash	black, porous, volcanic stone
habl	rope
hajar	stone
hajar al-aswad	hard, dense, black stone

hajar madawwar	curved faced stones for building the central column
hajar mathbih	roughly dressed stone used for the interior wall faces
hajl	a rhythmic work song
halal	permitted
hammam	bath(house)
haram	forbidden
harat	a quarter or neighbourhood of the city, usually named after the mosque there
hawsh	courtyard
hijra	the Hegira, the emigration of the Prophet Muhammad from Mecca to Medina in 622 AD.
hilal	brass or bronze crescent moon placed on the top of the minaret
hirawa	wood handle of a hammer or adze
hirz al-jinn	charm to protect against evil *djinn* (*jinn*)
hisba	market regulations
hizam manqush	decorative girth around the exterior face of a building
ihsan	doing what is beautiful, the third dimension of Islam
'ilm	knowledge of a religious, scientific, technical or trade type
iman	faith, the second dimension of Islam
islam	submission to God, the first dimension of Islam
jambiyyah	curved dagger worn by Yemeni men
jami'	Friday congregational mosque
ju'm	a hard black basalt stone
juss, jiss, or *guss*	gypsum-based plaster
kabal	thicker branches used as beams for supporting the staircase
khalta	mortar
khalta karee	mortar with stone aggregate
khalta safiya	fine mortar (without aggregate)
kurik	shovel
lahja	dialect
ma sha' Allah	"What God Wills"
madrassa	a (Qur'anic) school
mafraj	reception room, usually on the top storey of the house
mahram	inviolable place
makharif	residential districts or suburbs outside the city
makruh	reprehensible
mandub	recommended
maqshama	small, usually walled, garden for growing fruits and vegetables
ma'arifa	personal knowledge
mi'dhana	a minaret (derived from *'adhana*)

mihrab	prayer niche
mil'aqa	trowel
mitr	tape measure
mitraqa	hammer, chisel
mizan	level, plumb-line
muezzin	the person who makes the call to prayer
munawil	the labourer responsible for passing materials onto the *usta*
munshid	a singer who chants Qur'anic passages
muraba'a hajar	a square stone base
mushaqah	a daily wage
mushawath	trickster or magician
muta'allim	apprentice (recognised but not used by the Bayt al-Maswari)
nashad	wedding celebrations
naqsh	decoration
qaddad	a hard, water proof plaster applied to the exterior surfaces of buildings
qat	a leaf which contains mild stimulants that are released when chewed
quba muthallajah	exterior ribbed dome
muwaqqis	stone mason
nakhil	a sifter made of a metal mesh screening
qasabah	metal pole at the centre of the minaret column
qibla	direction of the Holy City of Mecca
qubbah	dome
qutb	axis, central column of the minaret
raqamah	to incise
raqama	incision patterns made on the *habash* stone
rassah	builder responsible for the infill between the inner and outer faces of the walls
rashidun	the rightly guided, referring to the first four Caliphs
rukn	corner
sab'	fingers, long thin branches which span between thicker wood beams used to support the floor or staircase
sabi al-usta	lad of the Master Builder (apprentice)
sadaqah	charitable endowments
sa'ila	dry river bed
sals	chevron patterning (in the brickwork)
saltah	a dish made from fenugreek, broth, rice and various other added ingredients
samsarah	*caravanseri* (turk.); hostel for traders located in the *suq*
saqa	to give water supply

251

saqifah	a shelter, or roofing
sauma'a (s.)	minaret
sawami' (pl.)	
sawm	courtyard of a mosque
sayyid	holy men, descendants of the Prophet
shaqi (s.) *shuqa* (pl.)	labourer
shalf	masonry rubble
shakush	hammer
sharaf	honour
sullam	ladder or staircase
suq	market
sum'a	reputation
ta'ban	tired
takhrim, 'aqd,	semi-circular fanlights in carved gypsum plaster &
or *gamariyyah*	stained glass
tamahhan	apprenticeship
tamreen	apprentice
tanzih	God's incomparability with anything else
taqlidi	traditional
tashbih	God's similarity to everything
tawhid	the Oneness of God
thana'a	second in command below the Master Builders
ujrat al-'ammar	(builder's) wages
usta (s.) *asatiyya* (pl.)	Master Builder
waqf	religious endowment
wara'	pious
za'al	braided goat's hair
zabur	mud courses
zajaj	glass
zakat	alms
zanbil	bucket, pail
zannah	long, ankle-length cotton shirt
zawwiyya	corner, nook, or a metal 90 angle
zubra	heavily weighted hammer

NOTES

Introduction: Sana'a, Craft and the Building Trade

1 Sponsored by the Canadian International Development Agency (CIDA) Awards for Canadians Programme.
2 Burrows, 1987:6.
3 Burrows, 1987:7.
4 Dresch, 1989:136.
5 1980:134.
6 Saqqaf, 1987.
7 Piepenberg, 1987.
8 Lewcock in Aga Khan, 1988:89.
9 Serjeant, 1980:135.
10 Al-Sabahi, H., 1996.
11 Bonnenfant, 1989.
12 Bonnenfant, 1989:162.
13 Costa in Serjeant, 180:157.
14 Messick, 1993:227–8.
15 Messick, 1978:400.
16 "Qat ... is a green leafy plant cultivated throughout eastern Africa and the Arabian peninsula, and containing two pharmacologically active ingredients, cathinine and cathine, the effects [from chewing] of which are something like amphetamine sulphate". Rushby correctly observes that, in Yemen, this leaf has "a pivotal role in poetry, music architecture, family relations, weddings and funerary rites, home furnishings, clothes, what people eat, when restaurants open and close, where roads go to and where not, who owns a car and who does not, office hours, television schedules, even whether couples have sex and how long it lasts". (Rushby, K. 1998:6–8).
17 Bonnenfant, 1989:102.
18 Bonnenfant, 1989:150.
19 Costa in Serjeant, 1980:158.
20 Messick, 1993:246–8.
21 For comprehensive discussions on Yemeni emigration and the impact of labour migration, see N. Abraham's (1983) article on the Yemeni immigrant community of Detroit; Friedlander's (1988) edited volume on Yemeni labourers in North America; Halliday's (1992) historical examination of Britain's Yemeni population; Carapico's (1988) study of the dependednce of the Yemeni economy on the export of labour to the oil-rich countries of the region; Myntti's (1984) study of

changes in the gendered division of labour as a result of male migration; and Weir's (1987) examination of the economic and social impact of labour migration on a Yemeni highland community.

22 Saqqaf, 1987.
23 1989:16.
24 Messick, 1978:109–110.
25 Weir, 1987.
26 Dresch, 1989:16.
27 op.cit.:21–22.
28 Messick, 1978:103.
29 Caton, 1990:223; Vom Bruck, 1991:75.
30 Burrows, 1987:8; Dresch, 1989:9.
31 Vom Bruck, 1991:72.
32 Messick, 1978:277–8.
33 See Vom Bruck, 1991:75–6.
34 See Saqqaf, 1989.
35 Vom Bruck, 1991: 75–6.
36 Costa in Serjeant, 1980:160–1.
37 Vom Bruck, 1991:72–3.
38 Saqqaf, 1989.
39 Bonnenfant 1989:229–31.
40 Lawless in Blake & Lawless eds., 1980:81.
41 Vom Bruck, 1991:74.
42 Messick, 1978:437–8.
43 See Kandiyoti, 1988.
44 Caton, 1990:222.
45 Carapico, 1998:107.
46 Ibn Khaldun, *The Muqaddimah: An Introduction to History.* Translated by F. Rosenthal. 1967:2:347.
47 Nasr, 1987a:246.
48 Burckhardt, 1976:196.
49 Nasr, 1987b:10.
50 Eickelman, D., 1989:308.
51 Nasr, 1987a:128–9.
52 See also Mermier, F. in Yemen Sanaa: Peuple Mediterraneens/ Mediterranean People, no. 46, 1989, for a discussion on the effect of the importation of goods on local craftsmen.
53 Lewcock, 1987:59.
54 Ibish in Lewcock, 1980:123.
55 In Piepenberg, in Saqqaf (ed), 1987.
56 Lewcock, R. 1986.
57 Dostal, W., in Serjeant & Lewcock, eds. 1983:263–5, 274.
58 Shelagh Weir, personal communication on April 3rd, 2000. Weir recommends that the government and concerned agencies take 'soft' measures which promote interest and pride in local craft, and which might stimulate demand.
59 For further discussion on these crafts see especially Bonnenfant, G. & P. Bonnenfant, 1981, and P. Bonnenfant 1987 and 1995.
60 Vom Bruck notes the recent denigration of the *sada* (the Prophet's progeny through 'Ali and Fatima responsible for recruiting the Imams, and who were the main targets of the Revolution) in contemporary politics as foreigners by hardline republicans. "Recalling the *sada*'s immigration to Yemen some ten centuries ago, they dispute their right to rule, or even to occupy a place in

Yemeni society, on the grounds that they lack legitimising Qahtani 'roots'."
However, "As the doctrinal dispute has become a controversy over authenticity,
the Zaydi *sada* who are prominent in the [Zaydi-oriented] Hizb al-Haqq [Party]
portray themselves as defenders of authentic indigenous knowledge." in Leveau,
Mermier and Steinbach (eds), 1999:181.

61 Note that the Imam was recognised by the Ottoman Empire as the religious and
secular leader of the Zaydi population by 1911.
62 Al-Iriyani, H., 1987.
63 Vom Bruck, G. in Leveau, Mermier and Steinbach (eds), 1999:174, footnote 14.
64 Mermier, F., 1997.
65 Al-Maytami, M., 1993; Unicef, 1993:145.
66 Unicef, 1993:151.
67 Unicef, 1993:146. An article in the *Yemen Times*, Sana'a March 9th through
15th 1998, vol. VIII issue no.10, page 1 & 7, reports similar statistics compiled
in 1994 by the population census. The *Statistical Yearbook* published in 1996
reports that 41.17% of children between the ages of six and fifteen are enrolled
in basic education, again with more than double the number of males to female
students, but registering a lesser difference in the urban to rural figures than the
previous statistical reports, recording 56% of urban children enrolled in basic
school as compared to just over 36% of rural children.
68 Statistical Yearbook, 1996:10. Like all statistics reported for Yemen, there are
conflicting figures for literacy, but all seem to indicate a less than 50% national
average. Al-Suwaida reports 41% (1995:95), and Unicef reports 33% (1993:12).
69 Dostal, W. in Serjeant & Lewcock, eds., 1983:274.
70 Beeston, A.F.L. in Serjeant & Lewcock, eds., 1983:36; and Serjeant in op.sit.,
1983:40.
71 Dostal, W. in Daum, 1987:361.
72 In Serjeant & Lewcock, eds., 1983:37.
73 A wax-like secretion from the sperm whale used in the fabrication of perfumes.
74 Ibn Khurdadbih in Sergeant, in Serjeant & Lewcock, eds., 1983:163.
75 In Macro, E. 1984:72.
76 Dostal, W. in Serjeant & Lewcock, eds., 1983:259.
77 Mermier, F. 1989:155.
78 Sergeant in Daum, 1987:165–6.
79 Unicef, 1993:22.
80 Lewcock, 1986; Marechaux, 1987.
81 Amadou-Mahtar M'Bow in Lewcock, 1986:8.
82 See Marchand, T., 2000.
83 Studio Quaroni in al-Sabahi, H., 1996:85.
84 Unicef, 1993:43.
85 In interview, May 8th 1997.
86 Several new tower houses were being erected during my stay in Sana'a, including
opulently palatial ones in the Talha quarter of the Old City and in Bawniyyah
near the Qa' al-Yahud.
87 As quoted in Piepenberg, in Saqqaf (ed), 1987:102.
88 Lewcock recommends in his report on 'Protecting and Conserving the Cultural
Heritage of the Old City' (in *The Old Walled City of San'a*) that certain controls
must be exercised over the construction of new building in the old city including
restraint on facade changes and on "the introduction of buildings made with
new materials or having novel elevational characteristics". (1986:109).
89 In discussion, June 5th 1997.
90 Beckman, B. in Ozdalga, E. & S. Persson, eds., 1997:3.

91 Carapico, S., 1998:2.
92 op.cit.:11.
93 Carapico, S., 1998:210–211.
94 Carapico, S., 1998:10–11.
95 Marchand, T., *Gidan Hausa*, unpublished report to the Canadian International Development Agency.
96 Ryle, G., 1949. *The Concept of Mind.*
97 Boyer, P., 1990. *Traditions as Truth and Communication.* Connerton, P., 1989. *How Societies Remember.* Jackson, M., 1989. *Paths Towards a Clearing.*
98 Fodor, J., 1983. *The Modularity of Mind.*
99 Jackendoff, R. & B. Landau, 1992.

Chapter One: The 'Addil Minaret

1 al-Hajari, 1942:59–60. Quote translated into English by T. Marchand & Shaif Jaralla.
2 Published in Arabic in Muhammad b. Ahmad al-Hajari, 1942: 71–73.
3 Vocke, H., 1973.
4 Serjeant, R.B., in R.B. Serjeant & R. Lewcock, eds. 1983: 317–321.
5 Sayyid Jamal al-Din, edited by 'Abdullah ibn Muhammad al-Hibshi, trans. by T. Mackintosh-Smith, 2000.
6 See Arabic version of poem in al-Hajari, op.cit.:73; and footnote 69 by T. Mackintosh-Smith in his translation of Sayyid Jamal al-Din, op.cit.
7 Vocke, H., op.cit. With regard to Qadi Abdullah al-'Arasi (ob. 1773), Zabarah's report that "all the *awqaf* of the Yemen were brought under him and he administered them well, dealing with them all in accordance with the *miswaddat* – this did not happen with anyone before him" (*Nashr al-'arf*, II, 152, in Serjeant & Lewcock, eds. 1983:153) would seem to contradict Vockes suggestion that al-'Arasi was corrupt like many of the inspectors of the *awqaf*.
8 R.B. Serjeant in Serjeant & Lewcock, eds., 1983:318–321. See Serjeant for a more complete commentary on the poem.
9 Lewcock, R.,1986:89.
10 Al-Hajari, 1942: 59–60.
11 Bloom, J., 1989:21.
12 Hillenbrand, R., 1996:129. "When the Muslims came to Madina they used to gather for prayer without any given summons to it; a lack which they discussed one day and some argued 'Let us have a bell like the Christians' and some said 'Let it be a trumpet like the horn of the Jews'. Umar said: 'Why not appoint a man to call the people to prayer?' And the Prophet said 'Rise, Bilal, and call the people to prayer'."
13 Serjeant in Serjeant & Lewcock, 1983:311.
14 Al-Hajari, 1942.
15 Bloom, J., 1989:191. See also Ettinghausen & Grabar; " ... the emphasis given to minarets in faraway lands ... demonstrate that for many centuries ... they served as a spectacular symbol of the presence of Islam.", 1987:37a.
16 Creswell, K.A.C., 1926:7. Creswell's theory refuted that of Thiersch (in *Pharos*) who supported the 'Pharos' theory of origin of the minaret initiated by A.J. Butler (in *Atheneum*, November 20, 1880, p.681). Thiersch argued that the octagonal Palestinian minaret came from Egypt. Creswell's challenge to Thiersch of an Ummayad origin of the minaret derived architecturally from the Syrian church towers was supported by Oleg Grabar.
17 Bloom, J., 1989:175–77.

18 Lewcock, R. in B.R. Pridham, ed., 1985:215. Also Serjeant & Lewcock. 1983:492.
19 Jean François Breton in Daum, W. ed. 1988:111.
20 In Serjeant & Lewcock. 1983:493.
21 Jamal al-Din, edited by 'Abdullah ibn Muhammad al-Hibshi, translated by T. Mackintosh Smith. 2000.
22 Lewcock in Pridham, ed., 1985:215.
23 See Bonnenfant for the Sana'ani prescription of house orientation. 1989:202.
24 Also see Hillenbrand 1996:134.
25 Bloom, J., 1989:175–177.
26 op.cit.:191.
27 Serjeant & Lewcock, 1983:311.
28 Hillenbrand, R., 1996:129.
29 See Madelung, W. in Daum, W., ed., 1988:175–6. Also, Finster, B. 1992:140.
30 1996:134.
31 Nasr, S.H., 1987:51.
32 Hillenbrand, 1996:133.
33 Bloom. J. 1989:191.
34 Hillenbrand, R., 1996:132.
35 Finster, B. 1992:134.
36 And he continues: " ... for it cannot be denied that the persistent equation of the minaret with the proclamation of power is hard to reconcile with its widely assumed role as a tower for the *adhan*" (1996:134).
37 1996:153.
38 Marcicq, A. and G. Wiet, 1959:65.
39 Lewcock, Serjeant & Smith, in Serjeant & Lewcock, eds. 1983:372.
40 Serjeant in Serjeant & Lewcock, 1983:321.
41 Serjeant, R.B., 'An early Zaydi manual of hisba', RSO, Roma, 1953, XXVIII:16.
42 Bloom, J., 1989:100 & 149. Also see Ettinghausen & Grabar, 1987:209.
43 Bloom, J. 1989:100, footnote.
44 Smith, R. in Serjeant & Lewcock, 1983:55.
45 op.cit.:56.
46 Finster, B., 1992:145.
47 See description of Zafar Dhi Bin minaret, Finster, 1992:132. Note that Wald claims that the al-Mansur Mosque at Dhi Bin was built as a funerary mosque by the first Rasulid ruler, Umar ibn Rasul, who assumed the title al-Mansur ('The Victorious') in 1229 at a time when Zaydi authority was restricted to the far north (Wald, P. 1996:123). The curved arcades of the facades in the mosque courtyard suggest an importation from the Maghrib via Egypt (Finster, B. 1992), and would therefore support Wald since the Umar ibn Rasul was originally the vizier of Saladin in Cairo. I would suggest that the finely decorated brick minaret, which stands as a separate structure and is stylistically distinct, would likely have been built at a later date than the mosque.
48 Lewcock also notes that the Zaydi prejudice against minarets does not seem to have persisted beyond the fifteenth century. (1986:89).
49 Al-Hajari, 1942:96.
50 Macro, E., 1984:70.
51 As quoted in Bidwell, R.L., in Serjeant & Lewcock (eds), 1983:118.
52 Al-Hajari, 1942. See respective chapters on each of the mosques cited.
53 Sayyid Jamal al-Din, edit. by al-Hibshi & translated into English by Mackintosh-Smith. 2000.
54 Finster, B. 1992:132.

55 Al-Hajari, 1942:121. Translated into English by T. Marchand, with assistance by Shaif Jaralla & Tim Mackintosh-Smith. The "whiteness and the brownness" would seemingly refer to the colours of the contrasting whitewash used to pick out the relief decorative patterns against the background of brown kiln-baked brick.
56 Serjeant, in Serjeant & Lewcock, eds. 1983:321 & 323.
57 Bonnenfant, 1989:188.
58 1996:72–73.
59 See also Serjeant & Lewcock, 1983:486–7, for a discussion on the use of the six-pointed star in Sanaʿa.
60 Bonnenfant, P., 1989:114.
61 Bloom, J., 1989:155–6.
62 Sayyid Jamal al-Din, edit. by al-Hibshi, trans. by Mackintosh-Smith. 2000.
63 In Serjeant & Lewcock, eds., 1983:45.
64 Bonnenfant. 1989:102.
65 See Serjeant & Lewcock, 1983:468 & 484, on the structural role of the masonry pier and wrapping staircase in tower house.
66 1996:134.
67 Hillenbrand, 1996:144.
68 op.cit.:146–7
69 op.cit.:148–152.
70 Lewcock in Daum, ed., 1988:211, footnote; and Hillenbrand, 1996:153 in reference to the Samanid Mausoleum at Bukhara.
71 In Serjeant & Lewcock, eds., 1983:362; and see Lewcock in Daum, ed., 1988:211.
72 1992:132.
73 Von Bothmer, in *Pantheon*, vol. 45, 1987; and in conversation.
74 Bloom, J., 1999:28–29.
75 Bloom, J., 1999:36–39; also Blair & Bloom, 1994:3.
76 Bloom, J., 1999:39. It is important to note that during the periods of Ottoman rule in Sanaʿa, the construction of minarets continued in 'Sanaʿani' style, following the precedent set by the minaret of the al-Madrasah Mosque.
77 See Bonnenfant, P., 1989:176.
78 In Serjeant & Lewcock, eds. 1983:362.
79 For example, see Hermann Burchardt's photograph of 'The souk in Sanaʿa', 1902 or 1907, which shows the decorated minaret of the ʿAqil Mosque on the lefthand side (Museum für Völkerkunde, SMPK, Berlin) reproduced in Dostal, in Daum (ed), 1987:363.
80 See Abdullah Zeid Ayssa's thorough investigation of *The Thermal Performance of Vernacular & Contemporary Houses in Sanaʿa, Yemen*. PhD thesis submitted to the Architectural Association Graduate School, London, 1995. Unpublished.
81 Lewcock, R. 1986:23–24. For further descriptions of the physical extent of the cathedral, see Lewcock, 1986:36.
82 Lewcock, R. 1986:25. During my own visits to this location, there was too much rubbish in the pit to discern the remains of any foundations.
83 In his study of *The Sacred Direction and City Structure* in Moroccan cities, Bonine concludes that, in some instances, the variations in *qibla* directions in a single city may indicate the influence of local topography. He also suggests that "perhaps slopes in the appropriate qibla direction were selected for the location of many of these medieval Islamic Moroccan cities". Bonine, M. 1990:70.
84 Nasr, S.H., 1987:44–45.
85 Hillenbrand, R., 1996:171.
86 Nasr, S.H., 1987:49–50.

Chapter Two: Foundations

1 Nasr, S.H., 1981:193.
2 Dresch, P. 1989; vom Bruck, G. 1991, in Leveau, Mermier & Steinbach (eds) 1999; Mundy, M. 1995; Weir, S. 1985, in Lawless (ed) 1987; Meneley, A. 1996; Stevenson, T.B. 1985; and Gerholm, T. 1980. Also see R. Lewcock, 1986, 'Social Structure and Way of Life', in his *The Old Walled City of San'a'*, p. 55–61.
3 vom Bruck, 1999b:154. Lewcock, however, states that the *muzayyin* (or *Bani Khumis*) class (and presumably the *akhdam*) often lived in separate quarters in some of the highland towns and in parts of the old city of Sana'a as well. (1986:56).
4 Lewcock, R. 1986:56.
5 Dostal, W. in Daum (ed) 1987:361. Note that the stratification in other parts of Yemen, including the rural highlands surrounding Sana'a, differs somewhat from the Sana'ani urban context, but the sada class tend to be consistently at the top of the hierarchy (see Dostal in Daum (ed), 1987).
6 Mundy, M., 1995:39. See also Dostal, in Daum (ed), 1987:350–353 for a description of the social stratification in the rural Yemeni highlands.
7 vom Bruck, G. in Leveau, Mermier & Steinbach. 1999:170.
8 Dostal, W. in Daum (ed), 1987:361–2.
9 vom Bruck, G. in Leveau. Mermier & Steinbach. 1999:170–171.
10 Dostal, W. in Daum (ed), 1987:362.
11 Lewcock, R. 1986:56.
12 Weir, S. in Lawless (ed), 1987:278.
13 Mundy, M., 1995:39.
14 Dostal, W. in Daum (ed), 1987:362.
15 The origin of the *Bani Khumis* group is apparently derived from a legend. See Mundy 1995:40–41 for a detailed account.
16 Dostal, W. in Daum (ed), 1987:362.
17 Mundy, M., 'Introduction' to Varanda. F., 1981:4.
18 Mundy, M., 1995:42.
19 Weir, S. in Lawless, R. (ed), 1987:291.
20 Mundy, M., 1995:40.
21 There were numerous other family relations of the Bayt al-Maswari residing in Sana'a. To the best of my knowledge, no others were involved in the building trade, but they were engaged in a variety of other trades and professions, including one who was a medical doctor.
22 My study concentrates exclusively on the male members of the Bayt al-Maswari. As a male ethnographer, I did not have access to the female members of the family. Even when invited into the homes of the builders, the women were never in my presence, and it would have been considered highly impolite (not to mention irrelevant since I was studying building, not domestic politics) for me to inquire about them.
23 Tapper, R. & N. Tapper, in Tapper, R. (ed), 1991:71.
24 Tapper, R. & N. Tapper, in Tapper, R. (ed), 1991:71–72.
25 Buchman, D. 1997:22.
26 The daily wages (*ujrat al-'ammar*) of builders were, however, regulated, like those of all trades, by the stipulations of the eighteenth-century *Qanun Sana'a*. (Bonnenfant, 1995:284)
27 Most probably after the revolution.
28 Ibish, Y. 1980.
29 Interview with Usta Sarhan ar-Rawdhi by Shaif Jaralla, February 1999.

30 Though the Bayt al-Maswari and Usta Sarhan al-Rawdhi reside in Sana'a, other builders that I worked with or interviewed, such as Usta al-Khowbani and Usta al-Turaygi, did not, but rather obtained commissions in Sana'a based on their reputations and specialisations in the building field. For instance, al-Turaygi is noted for his skilled construction of bathhouses.

31 Bukhari, 2:36.

32 Murata and Chittick suggest that *islam* pertains to acts, *iman* to thoughts and *ihsan* to intentions and that the three represent successive stages in the development of a Muslim (1996:xxxviii).

33 Maulana Muhammad Ali. *A Manual of Hadith*. publication date not stated:17.

34 Bloch critically remarks that despite Bourdieu's attempt to remedy structuralisms insufficient theories of learning which do not account for the "inevitably gradual construction of structured knowledge", his theory of learning *habitus* remains psychologically vague. (Bloch, M., 1998:18, footnote 4)

35 1989:191.

36 The siting of the second minaret I worked on, located in the southern suburb of Madinat al-Asbahi, was strategically chosen in the south-west corner of the mosque's large property in order to enhance views of the structure from considerable distances in all directions except true west which was blocked by a small mountain. The patron and builders originally intended to erect a fifty meter structure, but because of the site's proximity to the presidential palace, a maximum height of thirty-seven meters was stipulated by the national security forces. Further negotiations between the parties eventually resulted in a maximum height of forty-five meters.

37 M.E. Bonine also notes in his study of Moroccan cities that slope of the land is the prime factor in determining the orientation of religious buildings in the country's Islamic cities, and not the 'true' orientation toward Mecca (1990).

38 1983:361.

39 See Serjeant & Lewcock, 1983:468.

40 Translation by T. Mackintosh-Smith, 2000.

41 See Serjeant, R.B., 1976, for a more complete discussion of the significance of the ibex hunt and the use of animal horns in South Arabia.

42 See Serjeant and Lewcock, 1983:468.

43 From Hans Weir Dictionary of Modern Written Arabic, Third edition, 1976.

44 With reference to proportioning, Serjeant & Lewcock write that "Sana'ani builders prefer proportions derived from the perfect square, either singly or in multiples" when planning the length, width and height of rooms (1983:484).

45 The minaret of the Musa Mosque was added by the Imam al-Mansur al-Husayn in 1160 AH, who also at that time made repairs to the existing mosque which had been built in the 8th century AH (Hajj Muhammad bin Ahmed al-Hajari, 1942:121).

46 See Gunter Meyer in B.R. Pridham (ed.), 1985:147–164.

47 Mackintosh-Smith, T., 1997:100.

48 See Meyer in Pridham (ed.), 1973:161.

49 See Serjeant and Isma'il al-Akwa' for the Qanun's listing of builders' wages, in Searjeant and Lewcock, 1983:227.

50 In Serjeant and Lewcock, 1983:227, footnote 293.

51 Mackintosh-Smith, T., 1997:15.

52 See translation by Mackintosh-Smith, 2000.

53 See Serjeant and Isma'il al-Akwa' in Serjeant & Lewcock, eds., 1983:233–4.

54 Weir in Pridham (ed.), 1985:64.

55 Bloch, M., 1998:7.
56 Bloch, M., 1998:7.
57 In interview with 'Abdulwahhab al-Sayrafi.
58 Bourdieu, P., 1977:90.
59 Burckhardt, T., 1992:112.
60 Bourdieu, P., 1977:88–89.
61 Bloch, M., 1998:9.
62 Bloch, M., 1998:14.
63 Nasr, Sayyed Hossein, 1987:10.
64 Vom Bruck, G. 1994:164.
65 Op.sit:165.
66 Anne Meneley makes similar observations amongst the population she worked with in Zabid noting that the prevalent attitude is that there is no point in getting angry with a child less than five or six years old because they are not considered to have 'aql, or reason (1996:148).
67 Discussion of Vygotsky in J.A. Lucy, in *Rethinkinking Linguistic Relativity*, ed. Gumperz, John J. and Stephen Levinson. 1996:55–6. See also L.S. Vygotsky, *Mind in Society* 1978 (1930–34) and *Thought and Language* 1987 (1934).
68 See also Menely, A., 1996:163.
69 Asad, T., 1986:14. See also Messick, B., 1993.
70 See Marchand, T., 1994, for an analysis of the 'spatializing of time and practices' in the Carthusian Charterhouse.
71 See Eickelman, D.F. 1985:62.
72 1987:152.
73 1981:194/5.
74 1995:17.
75 Madelung, W. in Daum (ed), 1988:176.
76 vom Bruck, G. 1999b:154.
77 al-Iriyani, Hamid, in Daum (ed.), 1988:384–5.
78 vom Bruck, G. in Leveau, Mermier & Steinbach (eds), 1999a:174, footnote 14.
79 Tapper, R. & N. Tapper, in Tapper, R. (ed), 1991:72–73.
80 Statistics compiled between 1988 and 1992 on the educational sector of the country indicate that the highest percentage of graduates from the country's universities are from the falculties of 'education', as well as arts. *The National Report on the Situation of Women in the Republic of Yemen*, 1995:65.
81 *The National Report on the Situation of Women in the Republic of Yemen*, 1985:58. Of the 57.5% enrolled, only 27.5 % of these are females reflecting the estimated 80% illiteracy rate amongst women in the country. Op.cit.
82 Refer to religious text books published for the Yemeni basic school system, grades one to nine. Copyright of the Ministry of Education, 1995.
83 Murata and Chittick, 1996:9.
84 vom Bruck, G., 1999b:156. See also Eickelman, E. 1985.
85 vom Bruck, G., 1999b:157.
86 Op.cit.:177.
87 As a male ethnographer, I did not have occasion to meet or speak with mothers in my neighbourhood.
88 Dunbar, Charles F. in al-Suwaida (ed) 1995:64.
89 Buchman, D. 1997:21.
90 1987:10.
91 Foucault, M., 1977.
92 Kandiyoti, D., in Cornwall, A. and N. Lindisfarne (eds), 1994.
93 Eickelman, D. 1998:253.

94 See also Seyyed Hossein Nasr on Ibn Sina who is quoted on the topic of children's education as stating that "The great principle here is the inculcation of control of the emotions" (1987:152). See also Menely on the topic of Zabidi children and training emotions (1996:148).

95 The source of the word *hajl* is unknown. It's verb root, *hajala*, means "to cast amorous glances", or "to make sheep's eyes" (Hans Wehr, *A Dictionary of Modern Arabic*, 1980). Other words popularly used to describe work songs are *harajah*, whose second form verb root means "to joke or jest"; or *zaml*, whose verb roots are all associated with notions of camaraderie. *Hajl* was the only term used by the al-Maswari minaret builders.

96 The keen interest in my religious beliefs and attempts to convert me to Islam were short lived. In fact, I was always made to feel very welcome on the site by the builders despite any perceived or real differences, and on a number of occasions I was successfully protected by my colleagues from the inquisitive intrusions of potential troublemakers who thought it inappropriate that a non-believer should be working on a religious edifice.

97 Bourdieu, P., 1977:80.

98 1992:260.

99 Bourdieu, P., 1977:94.

100 In one instance, the client of a project which involved the conversion a former residence into a school and the addition of several stories to the existing structure, took an active role in the project management. Under normal circumstances, complete authority is vested in the hands of an *usta*. I worked as a labourer on this site for a brief period between minaret projects in order to compare and contrast the division of labour, the hierarchy of responsibilities, and the training process with that of the minaret.

101 Disciplinary techniques operate as an integrated system of "multiple, automatic, and anonymous power: for although surveillance rests on individuals, its functioning is that of a network of relations from top to bottom, from bottom to top, and *laterally*" (my italics). Foucault in Rabinow, 1984:192.

102 A devout son that has been absent for a period of time should, upon returning to his family home, kneel and kiss the knee of his father when he greets him. This is a sign of respect and confirmation of the gerontocratic order. Weir noted that when an important man came into the room of a *qat* gathering, "all present rose to kiss his knee and the hem of his gown in the traditional greeting of deference and respect". 1985:133.

103 *Yaaseen* is the title of sura number 36 of the Qur'an. Some theologians believe that *Yaaseen* is the name of the prophet, or that it was God displaying his power over the Arabic language. The meaning is unknown. This sura is normally recited when someone dies. One can also say "*yaseen 'alayak*" to someone in same context as 'God Bless you'.

104 *Haydarah* is an old word meaning 'lion', and therefore 'Ali is being compared with the virtues of a lion. This name was given to 'Ali, after courageously leading the *Jihad*.

105 In other words, 'Ali's strength is being compared to that of ten men.

106 In other words, 'Ali struck down the blasphemous Jews.

107 In other words, the sects of Islam.

108 The *thana'a*, Gayyid, believed that this verse, which was sung in the agricultural fields, refers to the workers calling forth the crops. It has also been suggested by others that this verse is more often sung in reference to passing along water from one person to another – *warad al-ma'*. It is a line sung in many *Hajl*.

109 A reference to the corner of the ka'ba stone called the 'Yemeni corner.'

110 Monday and Thursday are the two days of the week which Muhammad fasted outside the period of Ramadan and which are likewise recommended to all Muslims. Friday is the Holy Day of Islam.

111 Refers to the banner of Islam.

112 This refers to the doves which, along with a spider, protected Muhammad and Abu Bakir from soldiers by roosting at the entrance to a cave where they were hiding.

113 'Flower', 'sea', and 'by his name' all refer to the Prophet Muhammad.

114 1976:75.

115 For a more complete description of the building team and method which were responsible for the reconstruction of the defensive walls surrounding old Sana'a, see Marchand, T., 2000.

116 Bourdieu, P., 1977:87.

117 1987:5. My italics in the quotation.

Chapter Three: Making it Above Grade

1 From *Stephen Hero* as quoted in Grudin, 1990:58.

2 King, G. 1998:11.

3 See also Sergeant and Lewcock, 1983:481.

4 English translation by Mackintosh-Smith, T., 2000. Also see Sergeant and Lewcock, 1983:32.

5 Varisco, D., 1993:131.

6 op.cit.:132.

7 See english translation by Tim Mackintosh-Smith from the Arabic text edited by 'Abdullah ibn Muhammad al-Hibshi, 2000.

8 Varanda, F. 1981:101.

9 See also Bonnenfant, P. for a more complete discussion on the fabrication of clay bricks in Sana'a, 1995:317–321.

10 This is opposed to, what I contend to be, Bonnenfant's over-extended symbolic analysis of Sana'ani architectural decorative motifs. See Bonnenfant, P. 1989:188, 196.

11 See also Sergeant and Lewcock, 1983:472.

12 In the case of the al-Maswari minarets, the slit windows are aligned with the East and West, and the square base of the minarets are, in the vast majority of cases, oriented with the cardinal directions, as in the case of the two minarets which I laboured at, Masjid 'Addil and Masjid al-Ihsan.

13 1983:294.

14 Varanda, F., 1981:101.

15 Bloch, M., 1998:6.

16 There has been an enormous amount of study involving modular, or domain specific, theories of mind, first developed by J. Fodor, 1983, and taken up by numerous cognitive researchers. See Hirschfeld and Gelman, 1994.

17 1989:1.

18 op.cit.:7.

19 1980:116–122. Neither formal guild organisations nor initiation rituals pertained to the practices of contemporary traditional builders of Sana'a. See also Serjeant & Lewcock, 1983:479 footnote 30. Ibish compares Islamic Middle Eastern guilds to Sufi Orders (ibid:114), which are oppressed in Yemen, particularly in the predominantly Zaydi North. See my discussion about both guilds and Sufi orders in the introduction to chapter two.

20 op.cit.:116; Dilley in Coy, 1989:189.

21 Buechler in op.cit.:42; Deafenbaugh in op.cit.:170.
22 Coy, 1989:3.
23 Boyer, 1990:42.
24 Boyer, 1990:109.
25 Bloch, M., 1974.
26 Boyer, 1990:109; also see Sperber & Wilson, 1986:58.
27 Coy, 1989:124.
28 1995:239–40.
29 Herzfeld, 1995:241; Coy, 1989:2; Goody in Coy, 1989:247.
30 Goody in Coy, 1989:252.
31 1995:240–2.
32 in Coy, 1989:239.
33 As quoted in al-'Amri 1985:75.
34 Foucault, M., 1977. *Discipline and Punish: The Birth of the Prison.*
35 See M. Bloch for a discussion on intonation versus content in different forms of speech. 1974.
36 See P. Spencer for a discussion on the liminality of the Samburu Moran and the consequent production of anxiety, which serves to reinforce the existing social order. 1970.
37 1970:15.
38 op.cit.:22.
39 op.cit.:82.
40 Atran, S. in Hirshfeld & Gelman, 1994:335.
41 1996:151, my italics.
42 Note that Lewcock claims that "Yemeni men seemed to possess an innate colour sense. Bright colours were often worn in combination but almost never to poor effect". (1986:60).
43 See Gelman, Coley & Gottfried in Hirshfeld & Gelman, 1994.
44 op.cit.:344.
45 In Gumperz & Levinson (eds), 1996:46.
46 Maclagan, I. in Tapper, R. & S. Zubaida (eds), 1994:168.
47 In Gumperz & Levinson (eds), 1996:46.
48 All in Hirshfeld & Gelman, 1994.
49 Boyer in Gumperz & Levinson, 1996:208.
50 In Hirshfeld & Gelman, 1994:360.
51 1990:109.
52 Jackendoff, R. 1997:189–190.
53 Gumperz & Levinson in Gumperz & Levinson, 1996:32.
54 Keller & Keller in Gumperz & Levinson, 1996:117.
55 op.cit.:127.
56 op.cit.:119.
57 Bloch, M., 1991.
58 Ryle, G., 1949.
59 Fodor, J. 1983.
60 Keller & Keller in Gumperz & Levinson, 1996:115.
61 Taylor, C. 1985.
62 1996:118.
63 At the time of my study, the *muwaqqis* was earning 1400 riyals per day (about 11.25 American dollars) as compared with 500 riyals per day (about 4.00 dollars) for the labourers, including the 'apprentice(s)'. *Muwaqqis* Yahya Faraj complained, however, that he could earn up to 2000 riyals per day, an *usta*'s salary, at some other work sites.
64 Connerton, P. 1989; Bourdieu, P. 1977:94

65 1977:332.
66 Jackendoff, 1993:191.
67 1983:3.
68 op.cit.:7.
69 op.cit.:8.
70 op.cit.:13.
71 See Jackendoff on music, 1993:165–71.
72 1983:21.
73 op.cit.:35.
74 op.cit.:37.
75 See Jackendoff, 1993:51; Sperber and Wilson, 1994; Sperber, 1994; and McGinn in Warner & Szubka, 1994:114.
76 Fodor, 1983:41.
77 1992:43.
78 In Warner & Szubka, 1994:167–8.
79 Fodor, 1983:45/111.
80 1986:83.
81 1983:71–73.
82 op.cit:52.
83 Gumperz & Levinson, in Gumperz & Levinson (eds), 1996:22.
84 Jackendoff, 1993:159.
85 1986:176.
86 op.cit.:177.
87 1993:191.
88 1992:25.
89 op.cit.:43.
90 Flanagan, 1992:38.
91 1993:172.
92 op.cit.:186.
93 Sperber & Wilson, 1986:64.
94 op.cit.:230.
95 Jackendoff 1992:45.
96 Jackendoff & Landau, in 1992:99
97 Marr, D., 1982.
98 See Biederman, I., 1987.
99 Jackendoff & Landau, 1992:101–6.
100 op.cit.:120.
101 op.cit:112.
102 Ungerleider & Mishkin (1982) in op.cit:121
103 Heidegger, 1977:61.
104 Jackendoff & Landau, 1992:124.
105 Keller & Keller, 1999:25.
106 Keller & Keller, 1999:27.
107 Bloch, M., 1998:7.
108 Keller & Keller, 1999:7.
109 Keller & Keller, 1999:7.
110 Unpublished results from a cognitive experiment on object descriptions and spatial relations conducted by T. Marchand with students from the McGill University School of Architecture, 1995.
111 Keller & Keller in Gumperz & Levinson, 1996:119.
112 Studio Quaroni, in Piepenberg, F., in Saqqaf, A.Y. (ed), 1987:104. Piepenberg reports that "the establishment of a masonry school becomes inevitable, where

the traditional 'usta', the building master, can document and formulate his ancient building knowledge, passing it on to young students".

113 Lewcock, R., 1986:115. Lewcock recommends the "Establishment of mechanisms for the revival and training of all the traditional building crafts and skills needed for the efficient repair and maintenance of the old buildings". It should also be mentioned that during the period of Lewcock's study in the 1980s, there was a shortage of traditional builders because many young men had emigrated to the oil states for work, and many who remained in Yemen were lured to the then booming modernised building industries where they could earn higher wages. (1986:108). However, the Gulf War and return of the migrant workers (and the abrupt cut in remittances) has forced a reversal of those earlier trends noted by Lewcock. The probability remains that, with easing relations between Yemen and its neighbours, the exodus of fortune seekers may resume.

114 Bourdieu, P., 1977:94.

Chapter Four: Completing the Dome

1 Heidegger, M., 1992:320.
2 From Hans Weir Dictionary of Modern Written Arabic, Third Edition 1976.
3 op.cit.
4 op.cit.
5 From R.B. Serjeant & Isma'il al-Akwa', in Serjeant & Lewcock, 1983:227 footnote 293. Note that *sati* means *usta*.
6 Jamal al-Din 'Ali refers to the courtyards of Sana'ani mosques being paved with dressed slabs of *habash* stone in his *City of Divine and Earthly Joys*, translated by T. Mackintosh-Smith, 2000.
7 See Serjeant & Lewcock, 1983:319.
8 Tapper, R. in Tapper, R. & S. Zubaida (eds), 1994:225
9 Tapper, R. in Tapper, R. & S. Zubaida (eds), 1994:219–220.
10 *Al- Qur'an, A Contemporary Translation* by Ahmed Ali, 1994. sura 5: verse 90.
11 *Al Qur'an, A Contemporary Translation* by Ahmed Ali, 1994. sura 47: verse 15.
12 Note that Zaydis believe the Qur'an is a created work, whereas the Shafi'i Sunnis believe that the Qur'an is the word of God. (Lewcock, 1986:55).
13 Bonnenfant, 1995:284.
14 Coy, M. in Coy, 1989:119. Also see E. Goody, 1982, and M. Peil, 1970.
15 *Qaddad* is a mixture of ground black *hishash* stone and lime (*nurah*), mixed with water and left to ferment. It is applied painstakingly in layers and when completed, these are coated with one or more layers of plain lime which are successively polished with a piece of pumice. Finally animal fat or bone marrow from cows is rubbed into the surface to improve its weatherproof quality. Progress is slow and repetitive as only small patches of a surface can be adequately treated at a time. See Serjeant and Lewcock for a more complete description, 1983:479–480; or Bonnenfant's entire chapter dedicated to the topic, 1995:417–445.
16 Varisco, D.M. 1986:5.
17 Op.sit.:10.
18 The types of carved bricks have been outlined in the previous chapter. Their outward faces are carved as unit sections of diamond and chevron relief patterns.
19 In El-Said, Issam, and Ayse Parman. 1976:ix.
20 Nasr, S.H. 1987:243.

21 Costa, Paolo, & Ennio Vicario, *Yemen, Land of Builders*, 1977. Paul Bonnenfant has included a sufficient sampling of brick work motifs in his chapter on brickwork in *Sanaa: Architecture Domestique et Société*, 1995: 317–359.
22 Murata & Chittick. 1996:299–300.
23 Boyer, Pascal. 1990:42.
24 Ahmed, Akbar S. 1995:16.
25 Eickelman, D.F. 1985:57.
26 Vom Bruck, G. 1994:164.
27 Boyer, Pascal. 1990:100.
28 ibid:60
29 al-Sabahi, Hatim. 1996:118.
30 1992:47.
31 Laughlin, McManus, d'Aquili. 1992:103.
32 Priest on Brentano. 1991:116.
33 Heidegger. 1992:22.
34 op.cit.:37.
35 op.cit.:45–46.
36 McInerney. 1991:120.
37 Heidegger in Krell. 1977:61.
38 op.cit.:22.
39 McInerney. 1991:141–142.
40 op.cit.:135.
41 James in Flanagan. 1992:131.
42 Flanagan. 1992:41.
43 Klopf in Laughlin, McManus, d'Aquili. 1992:42.
44 Brodal, Dykes & Ruest in ibid:40.
45 in op.cit.:42. Also see Daniel Dennett, 1996:20–21.
46 Flanagan. 1992:38.
47 1992:114, my italics.
48 Dennett in Flanagan. 1992:42.
49 Laughlin, McManus & d'Aquili. 1992:119.
50 Flanagan. 1992:220.
51 Heidegger. 1992:106.
52 1996:196.
53 Cragg, K. 1956:151.
54 "How can you be ungratefultoward God when you were dead things and He gave you life? Then He will cause you to die, then He will give you life. Then you will be taken back to Him." (sura 2: verse 28). Murata & Chittick explain that "This verse lists the major stages that are normally discussed on the Origin and Return: nonexistence, this world, death, life in the grave, and the resurrection. After the resurrection, people will be divided into two groups, one of which enters the Fire and the other the Garden." (1996:196).
55 Ahmed Ali, *Al-Qur'an: a contemporary translation, sura* 48: verse 29. 1994:442.
56 Al-Bukhari, Sahih, *The Book of Prayers*, vol.1, book VIII, chapter 65.
57 *Al-Qur'an: a contemporary translation.* Ahmed Ali, 1994:173.
58 Blair & Bloom. 1995:70.
59 Ali, A., translator. 1994:260.
60 Blair & Bloom. 1995:70.
61 Meneley, A., 1996:67–68, footnote.
62 Serjeant, R.B. & Husayn al-'Amri in Serjeant & Lewcock. 1983:151.
63 op.cit.

64 op.cit.
65 1976:99.
66 In Serjeant & Lewcock. 1983:468.
67 Steele, J., in Serageldin & Steele. 1996:154.
68 Bloom, J. 1989:157.
69 Blair & Bloom. 1995:3.
70 1990:13.
71 Taylor, 1985:92.
72 1987:56.
73 1976:196.
74 See Burckhardt, 1976; and Nasr, 1987.
75 1994:133–134.
76 Date unknown:21–23, foot note. Also see Burckhardt, 1992:112–113.
77 1994:134–135.
78 Murata & Chittick. 1996:267.
79 Grudin, R. 1990:11.
80 Murata & Chittick. 1996:255.
81 1969:27–28. My parenthesis.
82 See Burckhardt, 1992:112–113.
83 al-Sabahi, H. 1996:119.
84 In conversation, 1997.
85 1990:116.

Conclusion

1 Vom Bruck, G. 1994:165. My italics for emphasis.
2 Eickelman, D.F., 1985:62.
3 op.cit.
4 Bourdieu, P., 1977:86.
5 Camelin, S. 1995:53.
6 Weir, S. 1985:130.
7 Eickelman, D.F., 1985:64.
8 Camelin, S. 1995:41–42.
9 Jackendoff & Landau, 1992.
10 See Bloch, M., 1991.
11 Carapico, S. 1998:210.
12 op.sit.:211.
13 Parodi, Rexford & Van Wie Davis, 1994:66 & 72.
14 Carapico, S. 1998:4.
15 Gellner cited in Ilkay Sunar, in Ozdalga & Persson, eds. 1997:9.
16 Carapico, S. 1998:11.
17 op.sit.:202.
18 op.sit.:60.
19 Beckman, B. in Ozdalga & Persson, eds. 1997:1–2.
20 Carapico, S. 1998:12.

BIBLIOGRAPHY

Abrahams, Nabeel. 1983. 'The Yemeni Immigrant Community of Detroit: background, emigration, and community life', in S.Y. Abraham & N. Abraham (eds.) *Arabs in the New World: studies on Arab-American Communities*. Detroit: Wayne State University, Center for Urban Studies. pp. 110–34.

Abu-Lughod, Janet. 1987. 'The Islamic City – Historic Myth, Islamic Essence, and Contemporary Relevance', in *International Journal of Middle East Studies*, 19:155–176.

Ahmed, Akbar S. 1988. *Discovering Islam: Making sense of Muslim history & society*. London: Routledge.

al-Amri, H. 'Abdullah. 1985. *The Yemen in the 18th and 19th Centuries: a political and intellectual history*. London: Ithaca Press.

al-Amri, H. & Serjeant, R.B., 1983. 'Administrative Organisation', in Serjeant, R.B. & Lewcock, R. (eds), *San'a': An Arabian Islamic City*. London: World of Islam Festival Trust. pp. 142–158.

al-Bukhari, S. No date. *The Book of Prayers, Vol. 1, Book VIII*.

al-Hajari, Muhammad b. Ahmad. 1942. *Masajid Sana'a*. Sana'a.

al-Iriyani, H. 1988. 'School and education: formation and development', in W. Daum (ed.), *Yemen: 3000 years of art and civilization in Arabia Felix*. Innsbruck, Austria: Pinguin-Verlag; Frankfurt am Main, Germany: Umschau-Verlag. pp. 375–88.

al-Maytami, M.A.W. 1993. 'Le marché du travail yéménite après l'unification', in *Revue du Monde Musulman et de la Méditerranée*, no. 67, pp. 121–9.

al-Sabahi, H.M. 1996. *Sana'a Yemen: Tradition & Modernity in Sanani Architecture*. Sana'a: funded by GTZ.

al-Suwaidi, J.S. (ed.). 1995. *The Yemeni War of 1994: causes and consequences*. London: Saqi Books.

Ali, Ahmed, 1994. *Al-Qur'an: A Contemporary Translation*. Princeton, N.J.: Princeton University Press.

Asad, T. 1986. 'The Idea of an Anthropology of Islam', in the *Occasional Paper Series* of the Center for Contemporary Arab Studies, Georgetown University.

As-Suwasa, Amat al-Alim, (ed.). 1994. *Democratic Development in Yemen*. Berlin: Klaus Schwarz Verlag.

Atran, S. 1994. 'Core Domains versus Scientific Theories: evidence from semantics and Itza-Maya folkbiology', in Hirschfeld & Gelman (eds), *Mapping the Mind*. Cambridge: CUP. pp. 316–340.

Ayssa, A.Z. 1995. *The Thermal Performance of Vernacular Contemporary Houses in Sana'a, Yemen*. Ph.D. thesis submitted to the Architectural Association, London, 1995.

Beckman, B. 1997. 'Explaining Democratization: Notes on the Concept of Civil Society', in Ozdalga & Persson (eds), *Civil Society Democracy and the Muslim World*. Istanbul: Numune Matbaasi, pp. 1–8.

Bel, J.M. 1988. *Architecture et Peuple du Yemen*. Paris: Conseil International de la langue Francaise.

Bender, B. (ed.). 1993. *Landscape: Politics and Gender*. Oxford: Berg Publishers.

Berlin, B. & P. Kay. 1969. *Basic Color Terms: their universality and evolution*. L.A.: University of California Press.

Bidwell, R. 1994. *Travellers in Arabia*. Lebanon: Garnet Publishing.

——, 1983. 'Western Accounts of San'a' 1510–1962', in Serjeant, R.B. & R. Lewcock, (eds) *San'a': an Arabian Islamic City*. London: World of Islam Festival Trust. pp. 108–121.

Biederman, I., 1987. 'Recognition-By-Components', in *Psychological Review*, 94 (2):115–147.

Blair, S., & J. Bloom. 1994. *The Art & Architecture of Islam 1250–1800*. New Haven: Yale University Press.

Bloch, Maurice. 1998. *How We Think They Think: anthropological approaches to cognition, memory, and literacy*. Oxford: Westview Press.

——, 1991. 'Language, Anthropology, and Cognitive Science', in *Man*, 26:183–198.

——, 1974. 'Symbols, Song, Dance and Features of Articulation: is religion an extreme form of traditional authority?', in *European Journal of Sociology*, 15:55–81.

Bloom, J. 1989. *Minaret: Symbol of Islam*. Oxford: Oxford University Press.

——, 1999. 'Revolution by the Ream: a history of paper', in *Aramco World*, vol. 50, no. 3:26–39.

Bonine, M. 1990. 'The Sacred Direction & City Structure: preliminary analysis of the Islamic Cities of Morocco', in *Muqarnas*, 7:50–72.

Bonnenfant, G., & P. Bonnennfant. 1987. *L'art du bois à Sanaa: architecture domestique*. Aix-en-Provence, France: Edisud.

——, 1981. *Les vitraux de Sanaa: premières recherches sur leur décors, leur symbolique et leur histoire*. Paris: Éditions du CNRS.

Bonnenfant, P., ed. 1995. *Sanaa architecture domestique et société*. Paris: Presses du CNRS.

——, 1989. *Les maisons tours de Sanaa*. Paris: Presses du CNRS.

Bosworth, C. E. 1996. *The new Islamic dynasties: a chronological and genealogical manual*. Edinburgh, Scotland: Edinburgh University Press.

Bothmer, H.C. Graf von. 1987. 'Architekturbilder im Koran: Eine Prachthandschrift der Umayyadenzeit aus dem Yemen' in *Pantheon*, vol. 45, pp. 4–20.

Bourdieu, P. 1990. *The Logic of Practice*. Oxford: Polity Press.

——, 1977. *Outline of a Theory of Practice*. Cambridge: University Press.

Boyer, P. 1994. 'Cognitive Constraints on Cultural Representations: natural ontologies and religious ideas', in Hirschfeld & Gelman (eds) *Mapping the Mind*. Cambridge: CUP. pp. 391–411.

——, 1990. *Traditions as Truth and Communication: a cognitive description of traditional discourse*. Cambridge: Cambridge University Press.

Buchman, D. 1997. 'The Underground Friends of God and Their Adversaries: a case study and survey of Sufism in contemporary Yemen', in *Yemen Update: bulletin of the American Institute for Yemeni Studies*, 39:21–24.

Buechler, H., 1989. 'Apprenticeship and Transmission of Knowledge in La Paz, Bolivia', in Coy, M. (ed) *Apprenticeship*. Albany: Suny Press, pp. 31–50.

Burckhardt, T., trans. from the German by W.Stoddart. 1992. *Fez: City of Islam*. Cambridge: The Islamic Text Society.

——, trans. by P. Hobson. 1976. *Art of Islam: Language & Meaning.* Westerham, England: World of Islam Festival Publishing.

——, trans. by D.M. Matheson. 1976. *An Introduction to Sufi Doctrine.* Wellingborough, England: Thorsons.

——, 1969. *Letters of a Sufi Master: the Shaikh al-'Arabi ad-Darqawi.* London: Perennial Books.

Burrows, R.D. 1987. *The Yemen Arab Republic – The politics of Development 1962–1986.* Boulder Colorado: Westview Press.

Camelin, S. 1995. 'Les pêcheurs de Shihr: transmission du savoir et identité sociale', in *Chroniques Yéménites,* vol. 3, pp. 38–56.

Carapico, S. 1996. 'Yemen between civility and civil war', in A.R. Norton (ed.), *Civil society in the Middle East,* vol. 2, pp. 287–316. Leiden, The Netherlands: Brill.

——, 1998. *Civil Society in Yemen: the political economy of activism in modern Arabia.* Cambridge: University Press.

——, 1988. 'Autonomy and Secondhand Oil Dependency of the Yemen Arab Republic', in *Arab Studies Quarterly,* vol. 10, no. 2. pp. 193–213.

Caton, Steve C. 1990. *Peaks of Yemen I Summon: Poetry as Cultural Practice in a North Yemeni Tribe.* Oxford: University of California Press.

——, 1986. 'Salam tahiyah: greetings from the highlands of Yemen', in *American Ethnologist,* vol. 13, pp. 290–308.

Cline, W. 1940. 'Proverbs and lullabies from southern Arabia', in *American Journal of Semitic Languages and Literatures,* vol. 57, no. 3, pp. 291–301.

Connerton, P. 1989. *How Societies Remember.* Cambridge: Cambridge University Press.

Cornwall, A. & N. Lindisfarne (eds) 1994. *Dislocating Masculinity.* London: Routledge.

Cosmides, L. & J. Tooby, 1994. 'Origins of Domain Specificity: the evolution of functional organisation', in Hirschfel & Gelman (eds) *Mapping the Mind.* Cambridge: CUP. pp. 85–116.

Costa, P.M. 1980. 'Sana'a', in R.B. Serjeant (ed.) *The Islamic city: selected papers from the Colloquium held at the Middle East Centre, Faculty of Oriental Studies, Cambridge, United Kingdom, from 19 to 23 July 1976.* Paris: UNESCO.

Costa, P.M., & E. Vicario. 1977. *Yemen: land of builders.* London: Academy Editions.

Coy, M.W. (ed), 1989. *Apprenticeship: from theory to method and back again.* Albany: State University of New York Press.

——, 1989. 'From Theory', 'To Method', and 'Being What We Pretend to Be', in M.W. Coy (ed) *Apprenticeship: from theory to method and back again.* Albany: State University of New York Press.

Cragg, K. 1956. *The Call of the Minaret.* New York: Oxford University Press.

Creswell, K.A.C. 1926. 'The Evolution of the Minaret: with special reference to Egypt', in *The Burlington Magazine,* March, May & June, 1926.

Damlugi, S. 1992. *The Valley of Mud Architecture: Shibam, Tarim, and Wadi Hadramut.* Reading: Garnet Publishing Ltd.

Daum, W. (ed.) 1987. *Yemen: 3000 Years of Art and Civilisation in Arabia Felix.* Innsbruck: Pinguin-Verlag.

Deafenbaugh, L. 1989. 'Hausa Weaving', in Coy, M. (ed.) *Apprenticeship.* Albany: Suny Press, pp. 163–180.

Dennett, D.C. 1996. *Kinds of Minds: towards an understanding of consciousness.* London: Weidenfeld & Nicolson.

——, 1991. *Consciousness Explained.* Toronto: Little, Brown & Company.

Dilley, R.M., 1989. 'Secrets and Skills', in Coy, M. (ed.), *Apprenticeship*. Albany: Suny Press. pp. 181–198.

Doe, D.B. 1983. *Monuments of South Arabia*. Santa Maria la Bruna, Italy: Falcon Press; Cambridge, England: Oleander Press.

Dorman, D., Dr. M. Aidarus, Dr. A. Hoskins, F. Salem, & M. Hamran, evaluation team. 1995. *Raymah Primary Health Care Project Evaluation 1995*. Yemen: International cooperation for Development.

Dorsky, S., T.B. Stevenson. 1995. 'Childhood and education in highland North Yemen', in E.W Fernea (ed.), *Children in the Muslim Middle East*. Austin, Texas: University of Texas Press. pp. 309–24.

Dostal, W. 1987. 'Traditional Economy & Society', in Daum (ed.) *Yemen: 3000 Years of Art and Civilisation in Arabia Felix*. Innsbruck: Pinguin-Verlag, pages 336–367.

Dresch, Paul. 1989. *Tribes Government and History in Yemen*. Oxford: Clarendon Press.

Eickelman, D.F. 1989, reprinted 1998. *The Middle East: an anthropological approach*. Englewood Cliffs, New Jersey: Prentice Hall.

——, 1985. *Knowledge & Power in Morocco: the education of a twentieth century notable*. Princeton: University Press.

El-Said, Issam & Ayse Parman. 1976. *Geometric Concepts in Islamic Art*. London: World of Islam Festival.

Ettinghausen, R. and O. Grabar. 1987. *The Art & Architecture of Islam 650–1250*. Harmondsworth: Penguin.

Evin, A. (ed.). 1983. *Development and urban metamorphosis: proceedings of Seminar eight in the series Architectural Transformations in the Islamic world, held in Sana'a, Yemen Arab Republic, May 25–30, 1983*. 2 vols. Singapore: Concept Media.

Fardon, R. 1995. *Counterworks*. London: Routledge.

Finster, B. 1992. 'An outline of the history of Islamic religious architecture in Yemen', in *Muqarnas*, 9:124–47.

Flanagan, Owen. 1992. *Consciousness Reconsidered*. Cambridge, Mass.: The MIT Press.

Fodor, Jerry. 1994. 'The Mind-Body Problem', in Warner, R. & T. Szubka, (eds). *The Mind-Body Problem*. Cambridge, Mass.: Blackwell Pub.

——, 1992. 'A Theory of the Child's Theory of Mind', in *Cognition*, 44:283–296.

——, 1983. *The Modularity of Mind*, Cambridge, Mass.: The MIT Press.

——, 1987. 'Modules, Frames, Fridgeons, Sleeping Dogs, and the Music of Spheres', in *Modularity in Knowledge Representation and Natural Language Understanding*, J. Garfield (ed). Cambridge, Mass.: The MIT Press.

Foucault, M. 1985. 'Other Spaces: the principles of heterotopia', in *Lotus International*. 48/49:9–17, 1985/86.

——, 1977. *Discipline & Punish: The Birth of The Prison*. Pantheon Books.

——, 1972. 'The Discourse on Language', appendix to *The Archaeology of Knowledge*. New York: Pantheon Books.

Friedlander, J. (ed.) 1988. *Sojourners and Settlers: the Yemeni immigrant experience*. Salt Lake City: University of Utah Press

Gell, A. 1992. *The Anthropology of Time: cultural constructions of temporal maps and images*. Lawrenceville: Princeton Academic Press.

Gelman, S., J. Coley & G. Gottfried. 1994. 'Essentialist Beliefs in Children: the acquisition of concepts and theories', in Hirschfeld & Gelman (eds) *Mapping the Mind*. Cambridge: CUP. pp. 341–366.

Gerholm, T. 1980. 'Knives & Sheaths: notes on a sexual idiom of social inequality in North Yemen', in *Ethnos* 1(3):82–91.

Golvin, L., M.C. Fromont, (et al.). 1984. *Thulâ: architecture et urbanisme d'une cité de haute montagne en République Arabe du Yémen.* Paris: Éditions Recherche sur les Civilisations.

Goodenough, W.H. 1990. 'Evolution of the Human Capacity for Beliefs', in *American Anthropologist,* 92:597–612.

Goody, E. 1989. 'Learning, Apprenticeship and the Division of Labour', in Coy, M. (ed) *Apprenticeship: from theory to method and back again.* Albany: SUNY Press.

——, 1982. *From Craft to Industry: the ethnography of proto-industrial cloth production.* Cambridge: Cambridge University Press.

Grandguillaume, G., F. Mermier, J.F. Troin, (eds.). 1995. *Sanaa hors les murs: une ville arabe contemporaine.* Tours, France: URBAMA, Université de Tours; Sana'a: Centre Français d'Études Yéménites. (Collection Villes du Monde Arabe, no. 1).

Grundin, R. 1990. *The Grace of Great Things: Creativity & Innovation.* New York: Tricknor & Fields.

Gumperz, J.J. & S.C. Levinson, (eds.). 1996. *Rethinking Linguistic Relativity.* Cambridge: University Press.

Halliday, Fred. 1992. *Arabs in Exile: Yemeni migrants in urban Britain.* London: Tauris.

Hansen, T., translated by J. McFarlane, K. McFarlane. 1964. *Arabia Felix: the Danish expedition of 1761–1767.* London: Collins.

Harris, W. B. 1985. *A journey through the Yemen and some general remarks upon that country.* London: Darf.

Hassid, S. 1939. *The Sultan's Turrets: a study of the origin & evolution of the minaret in Cairo.* Cairo: Imprimerie Misr.

Heidegger, Martin. 1992. *History of the Concept of Time.* Bloomington & Indianapolis: Indiana University Press.

——, David Farrell Krell, (ed). 1977. *Basic Writings.* San Francisco: Harper Collins Publishers.

Herzfeld, M. 1995. 'It Takes One to Know One: collective resentment and mutual recognition among Greeks in local and global contexts', in *Counterworks,* R. Fardon (ed). London: Routledge,

Hillenbrand, R. 1996. *Islamic Architecture: form, function and meaning.* Edinburgh: University Press.

Hirschfeld, L.A. & S.A. Gelman, 1994. *Mapping the Mind: Domain Specificity in Cognition & Culture.* Cambridge: C.U.P.

Hirschi, S., M. Hirschi. 1983. *L'architecture au Yémen du Nord.* Paris: Berger-Levrault.

Hobsbawm, Eric. 1983. 'Introduction: Inventing Traditions', in *The Invention of Tradition,* E. Hobsbawm & T. Ranger, (eds). Cambridge: Cambridge University Press.

Hume, David. 1975. 'Our Idea of Identity', 'Of Personal Identity' and 'Second Thoughts', in *Personal Identity,* Perry, (ed). California: University of California Press.

Ibish, Yusuf. 1980. 'Economic Institutions', in *The Islamic City,* R.B. Serjeant (ed). Paris: Presses Univ. de France.

Institut du Monde Arabe. 1987. *Sana'a: Parcours d'une cité d'Arabie.* Paris: Institut du Monde Arabe.

Jackendoff, Ray. 1997. *The Architecture of the Language Faculty.* London: MIT Press.

——, 1993. *Patterns In The Mind: language and human nature.* London: Harvester Wheatsheaf.

——, 1992. *Languages of the Mind: essays on mental representations*. Cambridge, Mass. The MIT Press.

Jackendoff, R. and B. Landau. 1992. 'Spatial Language and Spatial Cognition', in *Languages of the Mind*, R. Jackendoff, (ed). Cambridge, Mass.: MIT Press.

Jackson, Michael. 1989. *Paths Towards a Clearing*. Bloomington and Indianapolis: Indiana University Press.

——, 1983. 'Knowledge of the Body', in *Man*, 18:327–345.

Jones, W.I. 1936. 'Some Arab folk-tunes',in *Transactions of the Glasgow University Oriental Society*, vol. 7, pp. 10–16.

Kandiyoti, D. 1994. 'The Paradoxes of Masculinity: some thoughts on segregated societies', in Cornwall, A. & N. Lindisfarne (eds) *Dislocating Masculinity*. London: Routledge.

Kandiyoti, D. 1988. 'Bargaining with Patriarchy', in *Gender & Society*, 2/3:274–290.

Kennedy, J.G. 1987. *The flower of paradise: the institutionalized use of the drug qat in North Yemen*. Dordrecht, The Netherlands; Lancaster, England: Reidel.

Keller, C.M. & J. Dixon Keller, 1999. 'Imagery in Cultural Tradition and Innovation', in *Mind Culture, and Activity: an international journal*, 6 (1):3–32.

——, 1996. 'Imaging in Iron, or Thought is not Inner Speech', in Gumperz & Levinson (eds), *Rethinking Linguistic Relativity*. Cambridge: Cambridge University Press. pp. 115–129.

Ibn Khaldun, 1967. *The Muqaddimah: An Introduction to History*, 2nd edition, translated by F. Rosenthal. vol. 2:347. Princeton: Princeton University Press.

Kia, B., V.C. Williams. 1989. 'Saving Sana'a', in *Geographical Magazine*, vol. 61, no. 5, pp.32–36.

King, Geoffrey. 1998. *The Traditional Architecture of Saudi Arabia*. London: I.B. Tauris.

Laughlin, Charles, John McManus & Eugene d'Aquili. 1992. *Brain, Symbol & Experience: Towards a Neuro-phenomenology of Human Consciousness*. New York: Columbia University Press.

Lave, J. 1988. *Cognition in Practice*. Cambridge: Cambridge University Press.

Lawless, R.I. 1980. 'The Future of Historic Centres: Conservation or Redevelopment', in *The Changing Middle Eastern City*, G.H. Blake & R.I. Lawless (eds). London: Croom Helm.

Lawrence, D. and S. Low. 1990. 'The Built Environment and Spatial Form', in *Annual Review of Anthropology*, 19:453–505.

Leveau, R., F. Mermier and U. Steinbach (eds). 1999. *Le Yémen Contemporain*. Paris: Éditions Karthala.

Lewcock, Ronald. 1988. 'Working with the Past', in *Theories and Principles of Design*, Sevcenko (ed). Cambridge: The Aga Khan Program for Islamic Architecture.

——, 1986. *The old walled city of Sana*. Paris: UNESCO.

——, 1985. 'The Conservation of the Urban Architectural Heritage in Yemen', in B.R. Pridham, (ed). *Economy, Society and Culture in Contemporary Yemen*. London: Croom Helm.

Lewcock, R., R.B. Serjeant & R. Smith. 1983. 'The Smaller Mosques of San'a'', in Serjeant, R.B. & R. Lewcock, (eds) *San'a': an Arabian Islamic City*. London: World of Islam Festival Trust. pp. 351–390.

Locke, John. 1975. 'Of Identity and Diversity', in *Personal Identity*, Perry (ed). California: University of California Press.

Lucy, J. 1996. 'The Scope of Linguistic Relativity: an analysis and review of empirical research', in Gumperz & Levinson (eds), *Rethinking Linguistic Relativity*. Cambridge: Cambridge University Press. pp. 37–69.

Mackintosh-Smith, T. 1997. *Yemen: travels in dictionary land*. London: John Murray.

Maclagan, Ianthe. 1994. 'Food & Gender in a Yemeni Community', in Tapper, R. & S. Zubaida (eds), *Culinary Cultures of the Middle East*. London: I.B. Tauris. pp. 159–172.

Macnamara, J. 1994. 'The Mind-Body Problem and Contemporary Psychology', in Warner, R. & T. Szubka, (eds). Cambridge, Mass.: Blackwell Publishers.

Macro, E. 1984. 'Robert Finlay's journey in Yemen-1823', in *Proceedings of the seminar for Arabian Studies*, 14:67–76.

Madelung, W. 1988. 'Islam in Yemen', in Daum, W. (ed.), *Yemen: 3000 years of art and civilisation*. Innsbruck, Austria: Pinguin-Verlag. pp. 174–7.

Manzoni, R. 1884. *El Yémen: tre anni nell'Arabia felice: escursioni fatte dal settembre 1877 al marzo 1880*. Rome: Bota.

Marchand, Trevor. 2000. 'Walling Old Sana'a: reevaluating the resurrection of the city walls', in *Terra 2000 Preprints of the 8th International Conference on the Study & Conservation of Earthen Architecture*. London: James & James Scientific Publishers.

——, 1994. 'The Charterhouse: a vector of power and a container of memory', in *The McGill Review*, vol. 1:45–66.

——, 1992. Unpublished. *Gidan Hausa*, report to the Canadian International Development Agency, 176pp.

Marcicq, A., G. Wiet. 1959. *Le Minaret De Djam: La decouverte de la Capitale des Sultans Ghorides (X11e-X111e siecles)*. Paris: Librarie C. Klincksieck.

Marechaux, P. (ed.). 1987. *Sanaa: parcours d'une cité d'Arabie*. Paris: Institut du Monde Arabe.

Marr, David, 1982. *Vision*. San Francisco: Freeman.

Maulana Muhammad Ali. No Date. *A Manual of Hadith*. Lahore: M. Dost Mohammad.

McGinn, Colin. 1994. 'Can We Solve the Mind-Body Problem?', in Warner, R. & T. Szubka, (eds). *The Mind-Body Problem*. Cambridge, Mass.: Blackwell Publishers.

McInerney, Peter. 1991. *Time and Experience*. Philadelphia: Temple University Press.

Meneley, A. 1996. *Tournaments of value: sociability and hierarchy in a Yemeni town*. Toronto, Canada: University of Toronto Press.

Mermier, F. (ed.). 1997. 'Yémen: l'état face a la démocatie', in *Monde Arabe Maghreb Machrek*. no. 155, pp. 6–86.

——, 1989. 'Des Artisans Face aux Importateurs ou l'Ange Maudit du Souk', in *Peuple Méditerranéens*, no. 46, January-March 1989, pages 155–164.

Messick, B.M. 1993. *The calligraphic state: textual domination and history in a Muslim society*. Berkeley, California: University of California Press.

——, 1988. 'Kissing hands and knees: hegemony and hierarchy in shari'a discourse', in *Law and Society Review*, vol. 22, no. 4, pp. 637–59.

——, 1986. *Transactions in Ibb*. Ann Arbor: University Microfilms International. Ph.D. dissertation for Princeton University, 1978.

Meyer, G. 1985. 'Labour Emigration & Internal Migration in the Yemen Arab Republic', in B.R. Pridham, (ed). *Economy, Society and Culture in Contemporary Yemen*. London: Croom Helm.

Middle East Watch. 1992. 'Yemen: steps toward civil society', in *Newsletter/Middle East Watch*, vol. 4, no. 10. New York: Middle East Watch.

Mundy, M. 1995. *Domestic government: kinship, community and polity in North Yemen*. London; New York: Tauris.

Murata, S., W.C. Chittick. 1996. *The Vision of Islam*. London: I.B. Tauris.

Myntti, Cynthia, 1984. 'Yemeni Workers Abroad: the impact on women', in *MERIP Reports*, vol. 14, no. 5 (no. 124, June 1984), pp. 11–16.

Nagel, Thomas. 1994. 'Consciousness and Objective Reality', in Warner, R. & T. Szubka, (eds). *The Mind-Body Problem*. Cambridge, Mass.: Blackwell Publishers.

——, 1975. 'Brain Bisection and the Unity of Consciousness', in *Personal Identity*, Perry, (ed). California: University of California Press.

——, 1974. 'What is it like to be a Bat?', in *Philosophical Review*, 83:435–450.

Nasr, S.H. 1994 (first pub. 1966). *Ideals & Realities of Islam*. London: Aquarian.

——, 1987. *Islamic Art & Spirituality*. Ipswich, England: Golgonooza.

——, 1987. *Traditional Islam in the Modern World*. London: KPI ltd.

——, 1981. *Islamic Life & Thought*. London: Allen & Unwin.

Ozdalga, E. (ed.). 1997. *Civil Society, Democracy & the Muslim World*. Istanbul: Numune Matbaasi.

Parodi, C.A., E. Rexford, E. Van Wie Davis. 1994. 'The silent demise of democracy: the role of the Clinton administration in the 1994 Yemeni Civil War', in *Arab Studies Quarterly*, vol. 16, no. 4, pp. 65–76.

Peil, M. 1970. 'The Apprenticeship System in Accra', in *Africa* 40:137–150.

Perry, J. 1975. *Personal Identity*. Berkley: U. of California Press.

Piepenberg, F. 1987. 'Sana'a al-Qadeema: the challenges of modernization', in *The Middle East City*, Abdulaziz & Saqqaf (eds). New York: Paragon House.

Pridham, B.R. (ed.). 1985. *Economy, society and culture in contemporary Yemen*. London: Croom Helm.

Priest, Stephen. 1991. *Theories of the Mind*. London: Penguin

Rabinow, P. & G. Wright. 1982. 'Spatialization of Power: a discussion of the works of Michel Foucault', in *Skyline* March:14–17.

Ricoeur, Paul. 1971. 'Foreword' to *Hermeneutic Phenomenology: the philosophy of Paul Ricoeur*, Don Ihde. Evanston: Northwestern University Press.

Rushby, K. 1998. *Eating the Flowers of Paradise*. London: Constable.

Ryle, G. 1949. *The Concept of Mind*. Harmondsworth: Penguin.

Sapir, E. 1970 (first pub. 1921). *Language: an introduction to the study of speech*. Rupert Hart-Davis.

Saqqaf, A.Y. (ed.). 1987. *The Middle Eastern city: ancient traditions confront a modern world*. New York: Paragon House.

Sayyid Jamal al-Din 'Ali, edited by 'Abdullah ibn Muhammad al-Hibshi, translated by T. Mackintosh-Smith, 2000. *City of Divine & Earthly Joys: The Description of Sana'a*. Yemen Translation Series, Volume 3. Ardmore, PA: American Institute of Yemeni Studies.

Scott, H. 1947. *In the High Yemen*. London: John Murray.

Selvaratnam, V., O.M. Regal. 1991. *Higher education in the Republic of Yemen: the University of Sana'a*. Washington, D.C: World Bank. Policy, Research and External Affairs Working Papers, no. 676.

Serjeant, R.B. 1983. 'The Mosques of San'a': the Yemeni Islamic setting', in Serjeant, R.B. & R. Lewcock (eds) *San'a': an Arabian Islamic City*. London: World of Islam Festival Trust. pp. 310–322.

——, 1980. 'Social stratification in Arabia', in R.B. Serjeant,(ed.). *The Islamic city: selected papers from the colloquium held at the Middle East Centre, Faculty of Oriental Studies, Cambridge, United Kingdom, from 19 to 23 July, 1976*. Paris: UNESCO.

——, 1976. *South Arabian hunt*. London: Luzac.

——, 1953. 'An Early Zaydi Manual of Hisba', in *Rivista Degli Studi Orientali*, 28:1–34.

Serjeant, R.B. & R. Lewcock, eds. *San'a': an Arabian Islamic City.* London: World of Islam Festival Trust.

Sevcenko, Margaret Bentley, ed. 1988. *Theories and Priciples of Design in the Architecture of Islamic Societies.* Cambridge, Mass.: The Aga Khan Program for Islamic Architecture.

Shivtiel, A. 1996. 'Women in Arabic proverbs from Yemen', in *New Arabian Studies*, vol. 3, pp. 164–75.

Smith, R. 1983. 'The Early & Medieval History of San'a' ca. 622–1382/1515', in Serjeant, R.B. & R. Lewcock, eds. *San'a': an Arabian Islamic City.* London: World of Islam Festival Trust. pp. 49–67.

Spencer, P. 1970. 'Ritual in the Socialization of the Samburu Moran', in Mayer, P. (ed) *Socialization: the approach from social anthropology.* Tavistock ASA 8

Sperber, Dan. 1994. 'The Modularity of Thought and the Epidemiology of Representations', in *Mapping the Mind: domain Specificity in Cognition and Culture*, L. Hirschfeld & S. Gelman eds. Cambridge: Cambridge University Press.

——, 1985. 'Anthropology and Psychology: Towards an Epidemiology of Representations', in *Man*, 20:73–89.

——, 1975 (1991 edition). *Rethinking Symbolism.* Cambridge: Cambridge University Press.

Sperber, D. and D. Wilson. 1986. *Relevance: communication and cognition.* Oxford: Blackwell Publishers.

Stevenson, T.B. 1985. *Social Change in a Yemeni Highlands Town.* Salt Lake City: University of Utah Press.

Strawson, Galen. 1994. 'The Experiential and the Non-experiential', in Warner, R. & T. Szubka, (eds). *The Mind-Body Problem.* Cambridge, Mass.: Blackwell Publishers.

Swinburne, Richard. 1994. 'Body and Soul', in Warner, R. & T. Szubka (eds). *The Mind-Body Problem.* Cambridge, Mass.: Blackwell Publishers.

Tapper, Richard 1994. 'Blood, Wine & Water: social and symbolic aspects of drinks and drinking in the Islamic Middle East', in Tapper, R. & S. Zubaida (eds), *Culinary Cultures of the Middle East.* London: I.B. Tauris.

Tapper, Richard (ed) 1991. *Islam in Modern Turkey: religion, politics and literature in a secular state.* London: I.B. Tauris.

Tapper, R. & N. Tapper. 1991. 'Religion, Education & Continuity in a Provincial Town', in Tapper, Richard (ed) *Islam in Modern Turkey: religion, politics and literature in a secular state.* London: I.B. Tauris. pp. 56–83.

Tapper, R. & S. Zubaida (eds). 1994. *Culinary Cultures of the Middle East.* London: I.B. Tauris.

Taylor, Charles. 1985. 'Hegel's Philosophy of Mind', in his *Human Agency & Language: philosophical papers I.* Cambridge: Cambridge University Press. pp. 77–96.

Thomas, Julian. 1993. 'The Politics of Vision and the Archaeologies of Landscape', in *Landscape: politics and perspectives*, B. Bender, ed. Oxford: Berg Publishers.

Ungerleider, L.G. & M. Mishkin. 1982. 'Two Cortical Visual Systems', in Ingle, Gooddale & Mansfield (eds) *Analysis of Visual Behavior.* Cambridge, Mass.: MIT Press.

Van Hear, N. 1994. 'The socio-economic impact of the involuntary mass return to Yemen in 1990', in *Journal of Refugee Studies*, vol. 7, no. 1, pp. 18–38.

Varanda, F. 1981. *Art of building in Yemen.* London: Art and Archeology Research Papers; Cambridge, Massachusetts: MIT Press.

Varisco, D.M. 1993. 'The agricultural marker stars in Yemeni folklore', in *Asian Folklore Studies*, vol. 52, no. 1, pp. 119–42.

——, 1986. 'On the meaning of chewing: the significance of qat (Catha edulis) in the Yemen Arab Republic.' in *International Journal of Middle Eastern Studies*, vol. 18, no. 1, pp.1–13.

Vocke, H. 1973. 'Die Beschwerde der 'Addil Moschee', in *Zeitschrift der Deutschen Morgenlandischen Gesellschaft*, CXXIII.

Vom Bruck, G. 1999a. 'Being a Zaydi in the absence of an Imam: Doctrinal revisions, religious instruction, and the (re-) invention of ritual', in Leveau, R., F. Mermier and U. Steinbach (eds), *Le Yémen Contemporain*. Paris: Éditions Karthala. pp. 169–192.

——, 1999b. 'Ibrahim's Childhood', in *Middle Eastern Studies*, London: Frank Cass. vol. 35, no. 2:150–171.

——, 1994. 'Down-playing gender: Hatm rituals in Sana'a.', in *Quaderni di Studi Arabi*, vol. 12, pp. 161–82.

——, 1991 unpublished. *Descent and Religious Knowledge: 'Houses of Learning' in Modern Sana'a, Yemen Arab Republic.* Thesis for Ph.D., LSE, University of London.

Wald, P., translated by S. Wormell. 1996. *Yemen.* London: Pallas Guides.

Warner, Richard & Tadeusz Szubka, eds. 1994. *The Mind-Body Problem.* Cambridge, Ma: Blackwell Pub.

Watson, J. C. E. 1993. *A syntax of San'ani Arabic.* Wiesbaden, Germany: Harrassowitz.

Weir, S. 1987. 'Labour Migration and Key Aspects of its Economic and Social Impact on a Yemeni Highland Community', in R. Lawless (ed.) *The Middle Eastern Village: changing economic and social relations.* London: Croom Helm. pp. 273–96.

——, 1985. *Qat in Yemen: consumption and social change.* London: British Museum Publications.

——, 1985. 'Economic Aspects of the Qat Industry in Northwest Yemen', in B.R. Pridham (ed). *Economy, Society and Culture in Contemporary Yemen.* London: Croom Helm.

Werner, D. (ed.). 1988. *Yemen: 3000 years of art and civilisation.* Innsbruck, Austria: Pinguin-Verlag; Frankfurt am Main, Germany: Umschau-Verlag.

no author. no publisher. no date. *Forward–Looking Strategies & Policies for Health Development in the Republic of Yemen.*

no author. 1995. The National Report on the Situation of Women in the Republic of *Yemen.* Prepared by The National Prepatory Committee for the 4th International Conference for Women, Beijing, 1995.

no author. 1996. *Statistical Year Book.* Republic of Yemen, Ministry of Planning & Development Central Statistical Organization. Sana'a.

INDEX

Index of Topics

Index of Authors Quoted or Discussed